BEYOND AND BACK
Those Who Died and Lived to Tell It

BEYOND AND BACK

Those Who Died
and Lived to Tell It

by Ralph Wilkerson

Melodyland Productions
P.O. Box 6000,
Anaheim, Calif. 92806

Bible quotations are from the New American Standard Bible, ©
The Lockman Foundation, 1960, 1962, 1963, 1968, 1971, 1972,
1973, 1975 and are used by permission; also from the King
James Version.

First printing—June 1977

©1977 by Ralph Wilkerson

All Rights Reserved

Printed in the United States of America

Library of Congress Catalog Card Number: 77-72961
ISBN: 0-918818-00-1 (Softcover Trade) 0-918818-01-X (Hard-
cover)

DEDICATED TO
My Loving Wife, Allene
and to All the Others
Who Assisted Me
in Writing This Book

Introduction

This informative and inspirational book by Dr. Ralph Wilkerson answers many questions about the dimensions beyond our senses and mankind's apprehension and fear of them.

Of equal importance is the author's general survey of the Christian concept of death and judgment in contrast to certain occultic and cultic counterfeits.

The world is flooded with accounts of so-called "out-of-the-body experiences," reincarnation, and various versions of the beyond. Dr. Wilkerson marches boldly into the midst of this historic controversy with positive biblical answers expressed in language that the average person can understand and with which he can identify.

One of the best features of the book is its factual incidents confirming the survival of the human spirit after clinical death.

The miraculous, for the author, is not something that occurred in the dim, distant past; it is a reminder today that Jesus Christ is alive and that because He lives, we live also.

One may not agree with all of Dr. Wilkerson's conclusions, and it is not necessary to do so in order to appreciate his contributions. Here is a work which stems from the heart and ministry of a devoted pastor-evangelist and which is faithful in both its proclamation and defense of the Gospel of Christ. This book will evoke both praise and criticism, but it also will encourage faith in the authority of the Word of God, to which the author submits and by which he tests all human experience.

Dr. Walter Martin
Director, Christian Research Institute
Anaheim, California

All of us are fascinated about life after death. Dr. Ralph Wilkerson has stood along side the dead body of a man who came back to life. This account and others like it should make this book exciting reading.

Pat Robertson
President, Christian Broadcasting Network, Inc.
Virginia Beach, Virginia

Contents

Introduction

1

In Heaven for 30 Minutes

Universal interest has erupted on the topic of life after death. Major magazines have headlined stories about out-of-the-body experiences on their covers. Television shows on the phenomena have hit peak ratings. People are daring to deal with mortality and what follows.

It's a shocker! Everyone will come back to life someday. Regardless of your race, religion, age or social status, *you* will return. This mystery will unfold in the coming pages.

In the past no one wanted to lift the lid on the corpse, for the fear of death, as we will show in this book, is the father of all fears.

Society has tried to disguise death in many ways. The corpse has been draped with new phraseology. Graveyards are now memorial gardens. Markers have replaced tombstones. The undertaker is now called a director. Coffins have become caskets. The black hearse has become a pastel vehicle. An obituary is obsolete. Popular songs are sung instead of the great hymns of the church. A poem is substituted for prayer. Tributes are given, not sermons. Mausoleums, not Mother Earth, prevail because of land shortages. Dances and dinners have replaced mourning and black veils. In lieu of flowers, funds are sent to favorite charities. Often the corpse doesn't show up for the funeral; the remains have already been buried or cremated.

In this opening chapter I want you to meet a cabinet-maker who spent 30 minutes in a city in outer space and came back to Earth to tell about it.* National television shows and conventions have featured him as their guest. His encounter, one of the most fascinating on record, lifts the veil on the next world.

Round Trip to Heaven

On the morning of December 29, 1971 Marvin Ford suffered an acute, inferior myocardial infarction.

This cabinetmaker did not realize it at the time, but he was about to make a fantastic trip into the beyond, then return to enter into a ministry that would take him around the world.

Here is his story, as he tells it in *Voice,* an international magazine:

> In 1965 my wife Olive and I became affiliated with Christian Center (now Melodyland) in Anaheim, California. My wife played the piano for as many as five services each Sunday, and I directed the congregational singing.
>
> During this time I was promoted to foreman of one of

---

*Ford's full story can be read in his forthcoming book, *On the Other Side,* Logos International, Plainfield, N.J.

the largest millworks in Southern California, but my back was giving me a great deal of pain. I knew that one day I would no longer be able to continue the pace.

On December 13, 1971 I checked into the La Habra Community Hospital for a complete physical. I was put through every test imaginable.

Five days later my doctor dismissed me, explaining that I had some arthritis in my back. "But your heart," he said, "is like that of a 25-year-old man."

I returned to work. We spent a beautiful Christ-centered Christmas. Then on the morning of December 29, while transferring some Formica tops on a decline, the load slipped.

In desperation I attempted to steady it, but the strain was too much and something seemed to snap inside of me. At first I thought I was coming down with the flu and yet it felt as though an elephant had stepped on my chest.

The pain was excruciating, radiating down my left arm and up to my chin. I tried to raise my left arm. I couldn't. When my boss offered to drive me home, I stubbornly refused.

I laid out the work for about 20 men and then drove home by myself, breathing deeply in spite of the pain so that I might get as much oxygen to my lungs as possible.

My wife drove me to the hospital and within a few minutes several nurses were working over me. One put suction cups all over my chest to take an electrocardiogram. Another was trying to find my pulse. She couldn't.

They jerked the suction cups off and pushed my bed down to Intensive Care, started intravenous feeding with medication and gave me morphine for the pain.

My doctor arrived and set up a monitor above my head to

watch every heartbeat. He had pneumonia himself and I joked with him about who needed a doctor the most.

Then I offered prayer for him. The Lord touched him, and the next morning he was feeling fine. My condition was diagnosed as an acute, inferior myocardial infarction.

The following afternoon I went into cardiac fibrillation and arrhythmia (disturbance in the rhythm of the heart beat). A cardiologist was called, and I heard him tell my doctor, "Jim, there is no further need for me to return on this case. The patient cannot live through the night and probably for not more than two hours."

They thought I was unconscious and couldn't hear them talking.

In the meantime my wife was trying to get in touch with Pastor Ralph Wilkerson of Melodyland. He was in Texas, though expected back that day. About six o'clock in the evening he arrived at the hospital and was escorted into my room by the head security guard, who just "happened" to be a Christian.

I met tremendous opposition from the head nurse of Intensive Care Unit (ICU) when I first arrived at the hospital to pray for Marvin.

"I'm Pastor Wilkerson . . ." I began.

"Well, nobody's permitted in here," she answered shortly. "We're busy."

"I'm not here to visit," I answered patiently. "I'm here on a mission."

"Well, you're not coming in!"

"I'm sorry, but I *am* coming in."

"If you try, Reverend, I'll call the guard!"

"Go ahead and call the guard. I am a minister, and have a right to come in here. The family called me, so I am responding to a call. I'm sorry, but I'm going in!"

The nurse's antagonism was unbelievable.

"I'm calling the guard!" She stormed to the phone.

This much resistance is not normal. I sensed this was spiritual opposition, not just a stubborn nurse.

Nurses don't resist a preacher; they're trained not to react against ministers. Doctors and nurses in hospitals have been considerate in working with pastors, and I've attended seminars where hospital chaplains trained pastors on how to visit the sick.

At some point you have to be determined not to let people tell you what to do. You move on when you know it's a command of God. It was that determined faith that made me push ahead.

"What's the problem?" the guard asked, looking at the nurse and then at me as he approached.

"This minister is trying to force his way here, and I've told him he can't come in! This patient needs medical help right now. He doesn't need a pastor," she snapped.

Olive Ford, who was standing in the hall, offered to call her husband's doctor. The security guard, who is a Christian, walked with me to the head office to await word from the doctor.

"This is the most unusual thing I've ever had happen," he said as we walked down the hall.

A short time later the doctor notified the switchboard operator to have the guard escort me into ICU.

"Ralph Wilkerson's presence in that room is as important as mine, if not more important," Dr. J. David Rutherford said. I had met Dr. Rutherford on several occasions and his financial coordinator attends Melodyland.

"Let Reverend Wilkerson in," the guard commanded.

"Well, all right," the nurse grumbled. "But I think it's ridiculous!"

Death is an enemy. I could see the color of death on Marvin, and the smell of death was in the room. When I walked into the room, I could actually feel a spirit of opposition. As the nurses gave him oxygen, I prayed. I battled the opposition by quoting resurrection Bible verses such as the passage, "Greater is He that is in me than he that is in the world."

Marvin continues:

> There was great glory as he prayed for me, yet I continued to fail. My veins collapsed both Thursday and Friday.
>
> On New Year's Day my veins collapsed for the third time. I glanced at the clock on the wall near me. It was about ten minutes past two o'clock in the afternoon.
>
> Grim darkness entered my room. This had happened before when my veins collapsed. As it spread from the left side of my bed, the light began to vanish from the right.
>
> I was aware that death and life were waging warfare over my body. At times the light would gain; again this dismal darkness prevailed.
>
> Suddenly, all was as black as night. It was like the Carlsbad Caverns in New Mexico when tourists are taken far beneath the earth's surface and the lights are cut off momentarily.
>
> I felt my spirit leave my body like pulling the stopper out of a thermos bottle. Faster and faster my spirit traveled at an approximate 45-degree angle until ahead of me I began to see irridescent lights beckoning me into the most beautiful city I had ever seen.
>
> It was immense! I cannot verbalize the magnitude nor the splendor of that city. All was so overwhelmingly bright. It was like flying into Los Angeles on a smogless night at Christmas time when scintillating lights of every hue are arrayed in breath-taking magnificence. But this was magnified

into a brilliancy millions of times greater than anything this planet could produce.

I saw its gates of pearl, its streets of gold—not simply paved with gold—but a solid, yet crystal-clear gold.

I viewed its walls of jasper consisting of a heavenly green unlike anything I had ever seen before. I could see a marble-like vein running all through it. There were colors that are indescribable, for how can one describe a color without comparison to other colors? There is no comparison.

Then I saw one resplendent light emanating from the center of the right of God's throne. I recognized it as being Jesus, my Saviour!

Although I did not view Him as a corporeal being, He welcomed me into His presence, and I began to worship Him. I want to stress the dignity of worship there. There is nothing of the frivolity that we sometimes see in our earthly houses of worship. I cannot express its glory because it is infinity!

Then a strange thing happened. My attention was attracted to the hospital, and I could see my body lying upon the bed. A white car was parked near the entrance and a familiar person entered the building. I saw it all through those streets of gold! It was as though the hospital had no roof on it.

Before that time I had not known the layout of the hospital. As I continued watching, I saw Pastor Wilkerson enter, proceed down the corridor, turn to the Intensive Care Unit and knock on the door. . .

Earlier that day, I had been in my study preparing for Sunday's sermon. Suddenly I felt directed by the Holy Spirit to go to La Habra Community Hospital and pray for Marvin again. Heading straight for ICU, I did not meet opposition this time. The doctor had left orders to let me in anytime.

Mrs. Ford had just left, weary with hours of waiting and sleeplessness. The nurse let me into the room. I walked to the bed in the corner, pulling the curtain about us.

Here's how he later described the events that followed:

> Pastor Wilkerson picked up my left, lifeless hand and began to rebuke death in the all-authoritative name of Jesus, commanding my spirit to return to my body.
>
> There were three nurses in the room. Two were working over my body and the third was dialing the telephone. She could not contact a physician as most of them, I suppose, were either playing golf or enjoying the Rose Bowl game on New Year's Day. Even the house doctor was on an emergency call.
>
> I said, "Lord, it looks like trouble down there." I felt a tug earthward. Believe me, "The effectual, fervent prayer of a righteous man availeth much" (James 5:16).
>
> I asked Jesus what I should do. His answer was a question. "What do you want to do?" I replied that I preferred to stay with Him, but that if He would be with me and give me the fulfillment of a ministry I had never gained, I would elect to go back to Earth.
>
> This may sound like a fantasy, but every word of it is true. After some further dialogue, he gently chided me for doubting and gave me a command that I will never be able to discount: "Son, go back, and go with a commission—as an officer to join forces with those of my army who will usher in my Kingdom Age."
>
> I cannot explain all that is entailed in that statement. I only know that is what He told me.
>
> With that, my spirit returned to my body. It felt cold. I was later informed that this was due to shock. I smiled at one of the nurses, and she plugged in the cord to the monitor.
>
> Immediately I heard the familiar beep again. When I thought about looking at the clock, it was ten minutes until

three. I had been with Jesus at least half an hour of that time.

I had an immediate response from Marvin upon rebuking death. He not only smiled at the nurse, but said, "Oh, thank you, Pastor. Thank you for coming."

I could see he was having a spiritual experience, even though I didn't know what it was. Some days later I came back to visit him, and he began to tell me about a new awareness of reaching the world for Christ. I wondered how this carpenter could reach the world, then the Lord spoke to my heart, "I was a carpenter, and I reached the world."

Marvin told me particularly about South America and some other countries where the Lord was sending him.

He continues:

> For several days the glory of God flooded my hospital room. My physical eyes could not behold such radiance, and I kept them covered as much as possible with my hands. Whether I looked up, down, to the left or to the right, Jesus was there.

> The doctor said I was "clinically" dead. He wept and later told me he knew that they "lost" me. Even if I could have been brought back by mechanical devices, I should have been a human vegetable, for the brain cannot survive more than three or four minutes without oxygen and remain undamaged.[1]

Top Topic

Death is one of the five top topics of interest among U. S. high school students. For example, the neglected subject is getting special treatment in introductory psychology at Elk Grove High School in Illinois. And Rita Gold, a Miami Beach English teacher, recently startled her students by asking, "Suppose you could be told the exact time and mode of your death. Would you want to know the truth?" Also, many colleges and universities offering courses relating to death and the hereafter report these classes among the most popular.

Today people want to know as much about death as possible. Difficult as the subject is, today's generation is willing to come to grips with this life-related issue.

Solo Performance

Dying my death is one thing that no one else can do for me. Understandably, grief and disappointment surround death, but when the subject is treated biblically, it takes on a different perspective.

In this book you can encounter this awesome subject with a sense of peace as you become an explorer of eternity. On your journey beyond you will visit a city in which many questions will be answered, such as:

Do those in Heaven see what is going on down here?

Will I know my loved ones there?

Will earthly relationships such as husband, wife, parents and children extend in the beyond?

Will a baby remain an infant in Heaven?

Will the eternal city be large enough for everyone who's going there?

How long will we stay in Heaven, and what will we do?

Exactly *where* is Heaven in the universe?

Charles Panati, *Newsweek* science editor, says that due to modern resuscitation techniques, it is estimated that more than 1,000 people a year have encounters with death and come back to tell about it.[2] Thus the major part of this book will not deal with death, but with what is beyond.

During my years of pastorate and world travels, I have personally encountered many cases where people have died and lived to tell about it. I have not attempted merely to gather material on the subject. Nor have I tried to document the fact that people do come back from the dead. I am reporting firsthand experiences.

The Church Has More to Say

One of the foremost authorities on dying is Dr. Elisabeth Kübler-Ross, who has written numerous books. In *Questions and Answers on Death and Dying*, she wrote:

> I have specifically excluded chapters on "Religion and Life after Death" as well as chapters on "Bereavement and Grief." This was done not only because of lack of space, but because there are others who are more qualified to answer these questions.[3]

Just as Dr. Kübler-Ross does not write on the religious issues, I refrain from speaking on the medical implications. The church has more to say about the former issues than anyone else, since its greatest authority—Jesus—has been beyond and back.

The contemporary church is forced to face such perplexing questions as: Who should decide to pull the plug on life support machines—the church or the medical profession? Are mercy killings ever justifiable? What happens five minutes after you die? Should I have my body frozen at death in hopes of a discovered cure and eventual immortality?

In the coming chapters I offer some surprising answers to these fascinating questions. The psychological and spiritual changes in those who have been beyond and back also will be described as I point out some of the personality differences following their out-of-the-body experiences.

As you turn the page, you will meet a college president who died and came back with a message that has changed many lives.

Dr. Ralph Wilkerson, pastor-founder of Melodyland, and Marvin Ford, who spent 30 minutes in Heaven.

2
When Life Flowed In

A scene resembling a picture postcard could best describe that beautiful day in sunny Southern California. Nature had combined its forces of beauty in creating a colorful panoramic sweep of exotic flowers, blooming shrubs and waving palm trees silhouetted against a powder blue sky.

But sparkling skies almost gave way to desperate darkness that day when death tried to trip one of God's great servants.

"Pick up line one, Pastor . . . Edna Harrison is calling," my secretary said over the intercom.

"My husband has been rushed into intensive care . . . the doctors are not sure if he can pull through. We are at Hoag Memorial Hospital in Newport Beach."

"I'm leaving now. I'll hurry there as fast as I dare," I interrupted.

As I rushed from the office I told the receptionist, "Call the people on the prayer chain to pray for Dr. Harrison. He's had another coronary."

Plowing through the heavy freeway traffic, other thoughts flooded my mind as I gripped the steering wheel of the car.

Not Dr. Harrison. Dying? Irvine with a heart attack? Too much ahead of him—for the church—for his family—and maybe, selfishly, for me.

It was a miracle how he came to the little church in the women's club building. I'll never forget the unusual circumstance of his becoming part of our congregation.

At that time our attendance numbered less than 100, so when people came to visit, we smothered them with love. Our fellowship was small, but few slipped through without becoming part of us in those days.

Because of limited finances, I could do little advertising and follow up. So a few of the ladies were enlisted to write form letters to our visitors. Evidently a mistake was made, and the Harrison's younger son, Larry, filled out the visitor's card. So the letter was addressed to Larry instead of his parents. When Larry received it, he was so jubilant that he

Dr. and Mrs. Irvine J. Harrison and son, Larry, before his father's death.

insisted on his Mom and Dad driving the 25 miles one way through heavily congested traffic to church. The church that gets the children usually gets the parents, too.

We felt sure it was a courtesy visit when the Harrisons came that first Sunday. He had served as president of Southern California College in Costa Mesa before becoming the new director of National Educators Fellowship, representing Christian public school teachers, and kept a busy schedule. Too, they lived a long distance from the church. But the Harrisons came back.

During the next few years, Dr. Harrison and I shared many visions. We formed the School of the Bible for laymen, where today several thousand adults are enrolled. Melodyland School of Theology (MST) also was a goal we shared.

How could we possibly give him up?

I began to realize we were in a desperate spiritual battle. Pulling off to the side of the freeway, I began to resist Satan. I commanded him to leave Dr. Harrison alone.

At that moment God gave me a vision. It was compressed into three or four minutes, much like a dream, but I saw Dr. Harrison's funeral. I heard the songs, the eulogy and the sermon.

It was a coronation service. I was preaching, and there was joy. At one point his older son, Bob, was standing in front of the people, and everyone was marveling at his courage in eulogizing his father. I could see Mrs. Harrison, who had been so dependent on her husband, determining to be strong. I couldn't tell where the service was held, but there was a very large crowd there. (We had not yet purchased Melodyland.)

Instinctively I knew Satan was trying to kill Dr. Harrison. It wasn't his time to go . . . his work wasn't finished. I sensed there was a terrific battle.

While the Lord was speaking to me, God was giving other members of our congregation the same vision.

As this vision ended, I responded, "Yes, Lord, you have shown me the way it's going to be someday, but not now because I know this is not your plan."

At that moment I took dominion over Satan.

"I rebuke you, Satan, in the name of the Lord. This is not of God," I said with the power of the Spirit. "Lord, I am going in Your power."

As far as I was concerned, the battle had just been won. From that moment I was but a messenger of God's resurrection power.

I started up the car, and drove the rest of the way to the hospital.

Dr. Harrison told his part of the story in the May 1966 issue of *Abundant Life Magazine,* after he became professor of theology at Oral Roberts University. Here are excerpts from that account:

> May 28, 1965. A beautiful morning . . . and I was anxious to get going. Felt almost like I did in the days when I was just out of college, pioneering as a young evangelist . . . always raring to go for the Lord!

> Here I was again in pioneer work, this time doing student recruitment for a new university that was to open in September . . . Oral Roberts University (ORU).

> I felt the old spark in my soul and a peace in my heart, knowing that I was doing exactly what God wanted me to. I never felt better in all my life.

> In the morning I drove over to a church in Sacramento, California, to meet with a group of young people to talk with them about ORU. Then over to Stockton in the afternoon where I picked up my wife and son and drove on down to Merced, and preached that evening.

> At the close of the service in Merced, the pastor asked that I join him at the door to greet the people and to hand out

literature about ORU to those who wanted it.

Suddenly, out of the blue, I was stricken as I stood there shaking hands with the people. I developed a pain in my chest which rapidly became so severe that I asked my wife to take my place in the line while I went out to lie down in the car.

I had no idea what the trouble was. I hadn't been sick. In fact, I hadn't missed a day's work in the past 20 years.

Now whatever it was I had, was getting much worse until my arms and hands were immobilized, and I was perspiring like everything!

I began to pray and the only thing I could pray for was, "Lord, give me prayer partners."

That doesn't make sense, I thought. *I don't need prayer partners. Right now I need healing.* Again I started praying and all I could find were the words, "Lord, give me prayer partners."

"Well, this must be the Spirit of the Lord," I concluded and continued praying for prayer partners. After awhile the pain subsided, and I was back on my feet again.

The next morning my condition seemed to be all right and so we drove home—about 100 miles. As we traveled I couldn't help turning over in my mind something the devil had threatened me with when I first came to ORU, back in January. He had told me at that time, "You will see the glory of Oral Roberts University, but you will never enter therein."

I began to wonder if he was actually going to make his threat good. I knew that he would go all out to keep me from the greatest opportunity of my life . . . to teach young students in the Graduate School of Theology as well as the undergraduate division at ORU and help prepare them for their contribution to God and their generation. I rebuked him in the name of Jesus Christ, and went on my way praising the Lord.

Since the next day was Memorial Day, I waited till Monday to go to the doctor. "You have had a heart attack," he announced, "but I have no way of telling the extent of damage because it has now passed the acute stage. I recommend that you rest and cut down on all activities."

I was booked solid in Northern California for that week, and had planned to report in at Tulsa the Wednesday of the following week. But now I decided to take it easy and rest at home before returning to Oklahoma.

Several days later I suffered a coronary thrombosis, went into a state of shock, and was rushed to the hospital in an ambulance. The next morning I regained consciousness enough to see a doctor and two nurses at the foot of my bed.

"He is already becoming deoxygenated," I heard the doctor say, "we'll have to move him out of here right away." It was evident that he knew that the end had come for me.

I felt like my whole life had been a preparation for my faculty appointment at ORU. Only a few months to go until the opening of the university—and now it looked like the devil was about to make good his threat. It just didn't seem right.

Realizing it was the end of the road for me, the doctor called my wife and told her to summon the family. One by one they were allowed to come in my room to see me.

I remember especially my 13 year-old son, Larry, standing at the side of my bed. I commended him to God, and told him to take good care of his mother. He looked like a real soldier as he stood there. I knew he was ready to break but he didn't . . . I felt real proud of him.

Immediately after the last member of the family had left

the room, again I moved into a state where I was not aware of my surroundings . . . and then the most unusual thing happened:

> Suddenly, I was lifted up in my spirit, out of the body, and was looking down at my own body lying there. It looked like a great fish that had been washed up on the beach.

> Then I was aware of some beings in the background (I say "beings" because I don't know if they were persons or what). They were carrying on a conversation and I could hear one of them say, "He's dying." And another one said, "Yes, I see life oozing out of him."

> I became interested and wanted to see this, too, so I looked at this body and out of every pore was flowing a substance.

This happened about the same time I was driving on the freeway, and I sensed this spiritual battle. God was using me and others in the congregation to help Dr. Harrison in his fight for life. We were the prayer partners he had prayed for in Merced the week before.

His story continues:

> While these beings were discussing this scene, and I was looking on, there was an abrupt change in the atmosphere. I felt the room charged with faith.

> Something happened inside me and I spoke: "Why are they saying he is dying and that the life is flowing out of him? This is not right. This is not faith. Why doesn't someone say, 'Life, flow in' instead of 'out'?"

> I listened and none of them said anything. "If nobody else will say it, I'll say it then," and in the name of the Lord I commanded, "Life, flow in." As I watched that body, the substance that had been oozing out of the pores reversed itself and started flowing back into the body. Once again I became aware of my surroundings.

Within a period of time after that, I heard a commotion and looked up to see my pastor, Ralph Wilkerson, entering the room.

I took the elevator to the top floor of the hospital where the Intensive Care Unit was located. The Harrison family was in the waiting room, and I could tell they were very broken.

I didn't stop, but went right to ICU.

I met heavy resistance.

"You can't come in," the head nurse said sternly. She had a sheet draped over her arm and was disconnecting his life support apparatus.

"You don't belong in here. There's a *No Admission* sign on the door. Nobody is permitted in here now."

"But you don't understand," I argued. "I'm his pastor."

"No, you don't understand," she retorted. "It's too late; he's already gone. Go help his family."

She turned and began covering up his head.

Suddenly, I felt the Spirit of the Lord upon me. And in boldness I persisted. "I'm not here to visit! I haven't come to minister death. I came to minister life!"

I turned toward Dr. Harrison. There was no time to pray. "In the name of the Lord, I command life to flow into this body!"

Resurrection Bible verses began to pour from my soul.

"I am come that they may have life . . ." I began.

And every time I said the word *life* in those verses, it seemed another victory was won.

Soon I heard him groan and saw him move a bit.

I thought the nurse would faint. Frantically, she began to ring the bell.

A smile broke across his face. "Where am I?" he whispered.

For the next week I visited often with him. I took a tape recorder and placed it by his bed.

"Doc, anytime you have a little strength and are able to record, do it," I suggested. I was aware the experience was fresh on his mind, and God was giving him many insights.

He said later:

As I lay in the hospital, I began to reflect upon all that had happened. I saw more than my physical illness. I saw a spiritual battle, a conflict between the forces of Satan and the forces of God.

It came to me in sharp focus that we minister primarily on a spirit-to-spirit basis, not merely mind-to-spirit, or merely spirit-to-body. Christianity begins on a spiritual basis. It is a spiritual force dealing with the spirit, finding expression in the mind and the body to make man a whole person in Jesus Christ.

You see, in a few months I was to teach theology at ORU. I would be talking about things of the Spirit and relating them to the whole man.

Although I had been an evangelist, a pastor, a theologian, and for 13 years the president of a Christian college, I realized that I had not yet had all the preparation that God wanted me to have for this important new task.

He flashed back the scene of the miraculous day when His life flowed back into my body, and began to show me something about the three parts of man. Each part of man's body, mind and spirit has its own vocabulary and area of operation.

The nurse represented the physical part of man. The doctor represented the intellectual side of man. He had used his great skill and knowledge of my case and had come to the conclusion that I was dying.

In walked the pastor. He represents the spiritual aspect of man. God had called him to minister to my spirit. When he ministered in the Spirit, then the life of the Spirit of God flowed into my being. My mind and my body then responded, and I was made whole.

It is the Spirit that ministers to the spirit of men and women, and then the mind and body respond to this action. Thus the person is made whole.

God gave Dr. Harrison a fruitful ministry for the next eight years. A short time after his recovery, Dr. Harrison joined the staff of Oral Roberts University. Then just at the right time, at the beginning of Melodyland School of the Bible, God sent him back to us. After he returned to the Melodyland staff, Dr. Harrison began to lead teaching tours to the Bible lands.

He and Mrs. Harrison led many tours, but on his last one, for some reason, I told my wife, Allene, she must go along. I gave her explicit instructions on what to do in the event of an emergency.

The night before Allene was to leave, I was unable to sleep—a rarity for me. It is a family joke that often I'm snoring before I get completely into bed.

Awaking Allene I said, "I've been lying here thinking about Dr. Harrison. I've noticed the last few days that he seems fatigued. At the office, he barely gets up the steps at times. I just have an uneasy feeling about his health.

"If Doc should become ill, get the best medical help. Don't spare any expense, even the hospital if required. I have the strongest feeling he should not attempt this trip."

Then I blurted out my deepest fear, "And if he dies . . ."

"Oh, no, never . . . Ralph . . . please . . . we can't let a negative thought like that cross our mind," Allene exclaimed, springing straight up in bed, wide awake.

I had not wanted to distress the Harrisons, but could feel no peace about their travel. Compounding the added energies he had expended the last few weeks in caring for Edna following back surgery, plus other demands on his time, Dr.

Harrison was understandably fatigued. Gingerly, I had pointed out to Edna that her husband was exhausted and should not conduct a strenuous overseas tour. I also reminded Doc that Edna might find the trip extremely uncomfortable, because of her back problem.

Both of them had traveled extensively—he had made the Holy Land pilgrimage more than three dozen times—so it was not the lure of sight-seeing. Mike and Betty Esses were accompanying them and could lead the tour, I had mentioned. They need feel no pressure in changing plans.

But this couple (perhaps our greatest inspiration in those days) were unchanged in their eagerness to go. Though generally sagging physically at the moment, the Harrisons were dynamic and radiant in spirit. The trip would probably be a good recharge for them, and the refreshing would outweigh the restrictions, they reasoned.

The inevitable that was to happen I cannot question. I suspect that he, too, sensed that the grains of sand were slipping through the hourglass of life.

On Father's Day the Harrisons had the family over for a cookout. Dr. Harrison was barbecuing some steaks. His sons were asking about the tour. Then the subject got around to his health and whether he felt up to the long trip. Enthusiastically, he assured them he was looking forward to it and that he always gained new inspiration and immense fulfillment from the Bible land trips.

"In fact," he smiled to one of his sons while turning a steak on the grill, "I wouldn't even mind going to Heaven from the Holy Land someday."

The next day spirits were high, and everyone was waving good-bye as they boarded the plane at Los Angeles International Airport. That was to be my last glimpse of this cherished friend.

My wife was with the Harrisons almost constantly, watch-

ing them with careful concern. She recalls those final days like
this:

> Our first stop was Rome. It was the fourth of July, and
> we celebrated with a beautiful dinner at our hotel. Instead
> of fireworks the dining room was filled with a brilliant dis-
> play of popping flashbulbs. The atmosphere was festive.
>
> The next morning we loaded into the buses for a whole
> day of sightseeing. Edna saw us off, but stayed behind to
> rest.
>
> Though Doc had seen the sights dozens of times, he lis-
> tened to the guide as intently as if it were his first trip. Being
> a great historian as well as a theologian, he could always
> add some other dimensions that the guide didn't cover.
>
> Our last stop was at the catacombs. As the two busloads
> of us gathered around the guide inside the dimly lit sub-
> terranean caverns, there was hushed silence while we lis-
> tened. At the end of his talk, the guide remarked, "You
> know those early Christians believed that some day there
> would be a literal resurrection and their loved ones actually
> would rise up out of the graves!"
>
> "So do we," one of the young girls in our group shouted.
> Then somebody led out and all of us rousingly joined in
> singing *Faith of Our Fathers . . . we will be true to thee till
> death.* Our voices resounded throughout those corridors,
> bouncing off the musty dirt walls. Little did we know that
> this would be Dr. Harrison's final tour or that he would be
> singing his last great hymn of the church, which we later
> learned was a favorite of his.
>
> The next morning was free for shopping or rest, and in the
> late afternoon we headed for Athens. Again the seats were
> three across, and I sat with the Harrisons on the plane.
> Throughout most of the flight Dr. Harrison kept rubbing his
> chest.

Earlier in the day while shopping in Rome he had stopped abruptly several times with severe chest pains and shortness of breath. I had expressed concern, but both of them related how their physician had explained that remarkably new tissue was growing in the damaged area of his heart and that some discomfort was to be expected with the changes that were going on.

When arriving at our hotel in Athens, most of us went straight to our rooms as we were beginning to feel the jet lag. About one o'clock in the morning I was awakened by someone pounding on my door.

"Allene! It's Betty!" I heard through the door. Though calm and controlled, the voice signaled trouble. Bursting the door open, I had a sickening sensation that it was Dr. Harrison, even before she said a word.

"Dr. Harrison has had a heart attack. He's dying! Mike is with him. . . ."

Dr. Mike Esses, later dean of our laymen's school, was assisting Dr. Harrison with the tour. He gives this account of the event:

About one in the morning, the shrill ring of the telephone brings Betty staggering to her feet. In a fog, I hear her answer the phone, and then I hear alarm come into her voice. Betty shakes me wide awake. I hear her say, "That was Edna! Irvine must be having a heart attack!" As I run out the door, I yell at Betty to start praying.

Taking the stairs two at a time, I feel panic begin to rise and take over. "Lord, God, stand with me now so I can help" is my prayer as I approach the Harrison's room.

There is an instant response to my knock, and Edna shows me to the bathroom where Irvine is sitting. There is no doubt that he is in terrible trouble.

I telephoned Betty and told her to call Dr. Don Brandenburg, who is on the tour with us, and Don Pierotti. "Get them here as fast as you can!" I can hear myself yelling. Betty says, "Okay, Mike, take it easy. Everything is in God's hands."

Her steadiness has a calming effect on me, and I turn to Irvine as I hang up the phone. His face is etched with pain, his breath coming in short gasps. He is holding his chest, barely able to speak. I lean over him and ask, "Irvine, have you taken your nitroglycerine?" I hear a barely discernible *"No."*

Edna gets his pill for me, and Irvine manages to get it down. I sit back on my haunches, holding his hand, waiting for the pill to take effect.

In a matter of minutes, there comes one last audible breath, and Irvine falls over into my arms. I don't know whether he is unconscious or dead. At that moment, Dr. Brandenburg and Don Pierotti burst into the room.

The three of us get Dr. Irvine Harrison stretched out on the hard bathroom floor. Don Pierotti gives mouth-to-mouth resuscitation, and Dr. Brandenburg massages Irvine's chest.

Betty arrives, and Dr. Brandenburg tells her to go to the lobby to get medical help. "We have to have adrenalin," he tells her. On her way, Betty stops and tells Allene Wilkerson what's happening. Allene goes to be with Edna, Dr. Harrison's wife.

In the lobby, Betty finds no one around who knows what to do. The manager calls for an ambulance, but he says it will probably take a while for it to arrive.

In the Harrison's room, Dr. Brandenburg is keeping Irvine's heart beating by pressing rhythmically on his chest, and Don Pierotti continues to give mouth-to-mouth resuscitation.

Both men are dripping with perspiration. Their knees are in agony. They have been kneeling on the marble bathroom floor for nearly an hour.

Mahlon McCoury and I have been praying, reading the Psalms aloud, crying out to God to perform a miracle. Everyone on the tour has been interceding for Dr. Harrison.

I look up from my Bible to see Betty come through the door. The men with the ambulance are on their way up by the stairs, as the European elevators are not large enough to accommodate the men and their stretcher.

As I watch the Greek ambulance men and Dr. Brandenburg work to get Irvine on the stretcher, I realize these are almost battlefield conditions compared to what we have come to expect at home when there is an emergency.

The stretcher is made of two poles with a piece of canvas slung between them. There is no adrenalin, no resuscitator, just a small bottle of oxygen and a mask that they try to put on Irvine's face.

Dr. Brandenburg tries to explain that the oxygen will do no good, because Irvine isn't breathing. The men grab the ends of the stretcher and start down the stairs. At each landing, Dr. Brandenburg makes them put Irvine down for a moment, while he massages Irvine's chest and Don breathes for him.

We finally get the stretcher into the ambulance. Dr. Brandenburg and Don get inside. I turn to Mahlon to tell him to watch over the group; then I get in a cab and follow the ambulance to the hospital.

When we arrive, we find them taking Irvine into the emergency room. The Greek doctors have taken over. They are hooking Irvine up to an EKG machine, and the doctor calls for adrenalin. I watch the scope on the heart machine. There is just a straight line across the middle of it. No heart activity.

The doctor shoots the adrenalin into the heart; then he begins to pound on Irvine's chest. As long as he pounds, the lines on the scope show some peaks, but as soon as he stops, the line becomes straight again. They repeat this procedure, but with no more success than the first time.

Finally the doctor looks up and shrugs his shoulders in the universal gesture of "It's no use." I watch him as he folds away his instruments. After two hours of the most intense endeavor, the struggle is over.

I can't accept this finality. I take the authority granted to me by the Lord Jesus Christ. I hear myself saying, with a boldness that I don't possess, "In the name of Jesus Christ, Irvine, I command your spirit to come back to your body!"

The stillness in the room is a hushed expectancy; I know that God is going to move. I can feel him all around me. I just have to believe enough to see this miracle.

In an even louder, more bold voice, in a triumph of belief that comes from the very depth of my soul, I shout, "Father, in the name of Your Son, Jesus Christ, I command this spirit to come back into this body!"

There is movement. I know that Irvine is here again.

I can't have been mistaken. With my own heart pounding like it is going to leap from my chest, I lean closer and gaze upon Irvine's face.

There is a faint flutter of his eyelids, then slowly, ever so slowly, his eyes open. He moves his lips into a faint smile. I notice his eyes are different. They have beheld God, and in an instant, even before the slight shake of his head back and forth indicating *no,* I know that Irvine doesn't want to come back.

Yes, Irvine is healed in the most perfect healing of all. He is where we all want to be. Where there is love, peace, joy

— and with the Lord that he has loved his whole life. This is his day of triumph.

I stand by Irvine's side as he peacefully goes to be with his Maker, and I rejoice for him . . .[1]

A college-age girl was standing in the doorway with Allene when Dr. Harrison moved his head and then went beyond. "Did you see that?" she gasped in astonishment. Turning to each other, they stared in wonder, unable to frame into words the picture they had just witnessed. Allene continues the story from her perspective:

It was incredible beyond belief but vividly apparent that Dr. Harrison did not want to come back — something indescribably magnetic was pulling heavenward.

Back in the lobby, Mrs. Harrison found a telephone and placed a call to her older son, Bob. She located him in San Francisco where he and his wife were attending a Full Gosped Business Men's convention. As the telephone was ringing she said to me, "Oh, I just remembered . . . today is Bob's birthday."

Instead of wishing him a happy birthday, she had to tell him that he had lost his father.

But the comfort of the Holy Spirit was already with Bob. After the convention session some of their friends wanted Bob and his wife to go out for dinner. Somehow he felt the need to be alone in order to fast and pray.

Bob was on his knees in prayer and fasting when his mother's voice came on the telephone. Immediately he began to pour strength and comfort into his mother. At the end of their conversation, Edna handed me the phone. I could not believe it was Bob. He was at peace, knowing that "all things work together . . ." and providentially this, too, had to be God's timing.

On the plane from Athens to Los Angeles, Edna finalized
her plans for the service. It would be a coronation — not a
funeral. Irvine had been promoted and deserved a royal
homegoing. When the family was reunited in Los Angeles,
Bob showed his mother the plans for a *coronation*. Almost
every detail of the music, tributes and persons asked to par-
ticipate matched her plans.

That same funeral service that I had seen eight years
before took place.

In the years that have followed, God has become Edna's
strength, and she conducts a program in her spacious home
called New Lease. A hundred or more gather each month to
give encouragement and strength to those who have lost a
loved one.

The most amazing thing that happened to Dr. Harrison
after his first resurrection was not that he came back to life.
It was the change in his personality. His values changed and
he was more sensitive to the Holy Spirit.

Dr. Harrison gave this account of the change in the magazine article:

> I entered into a deeper relationship with the Lord. I knew
> that I could have gone home to Glory, but I am here because
> of Him.

> When you stand on the very brink, look over, and see
> yourself moving from time to eternity, suddenly you realize
> that nothing else matters — in fact, there just *is* nothing else
> — save Jesus.

> God established this goal in me: To live for, look for,
> move in, and anticipate continually the move of the Holy
> Spirit in my spirit, so that I might communicate the life of
> the Spirit to those with whom I come in contact daily.

Unknown to all of us, Dr. Esses later was to have his

encounter with death. I remember the intense spiritual struggle at the hospital as the forces of darkness battled to keep
me from entering his room.

In the next chapter I will tell how I recognized the spirit
of death and overcame it by the power of the Holy Spirit.

3
It Wasn't His Time to Go

A hefty nurse, already exasperated with ministers coming and going in the Intensive Care Unit, met me at the door of the ward. She and others were frantically trying to revive Dr. Michael Esses,[1] who had suffered a coronary. Certainly she wasn't in a mood for another preacher.

"I'm sorry. It's too late," she said gruffly. "There's nothing you can do."

The door was slightly ajar.

"I'm a pastor," I argued.

Let me say here that so often doctors and nurses have the wrong idea about the ministry.

As a minister I am a servant of the people, and I was called there by the family. The sick person doesn't work for the hospital, the hospital works for the sick person. My mission is just as important, maybe more important, than the physician's. I'm not there to visit, and the hospital's code on visitors is to be respected. Too often ministers have been pushed aside because they have been apologetic when they walk into a room to pray for people. At times I've asked the doctor to wait a moment while I pray for the patient. In almost every case the physician has been cooperative.

It seemed when Mike was in the battle for his life I encountered Satanic power. So I determined to persevere in the Spirit and break through.

Satan hates resurrection. He knew if he could prevent Jesus' resurrection, there would be no church, and the preaching of the cross would be in vain.

Ultimately, all God's children will be resurrected. And it is a testimony to God's faithfulness if some are resurrected now. It says to the world that God is greater than death.

In my years of hospital visitation, the only times I have encountered difficulties in seeing a patient is when God intended to perform a resurrection miracle.

When the Lord prayed for Lazarus, the Bible says, "Jesus groaned (or snorted like an angry stallion) in the spirit. In most of the biblical cases great conflict or agitation with Satan occurred.

Hebrews 2:14, 15 says:

"... that through death he might destroy him that had the power of death, that is, the devil; and deliver them who through fear of death were all their lifetime subject to bondage."

I could see the head nurse was in no mood for any kind of logic, so I pushed the door open and walked in.

ICU at Anaheim Memorial Hospital was a large room with about 10 beds. When I saw Mike I knew the spirit of death had come, but it didn't frighten me because I knew Mike belonged to God, and that it wasn't his time to go.

Many times Dr. Esses has told the story of the miracle God performed in Intensive Care that afternoon. Here are excerpts from his accounts given during a recent regional convention and in his book, *Michael, Michael, Why Do You Hate Me?*

I was a hot shot. I thought I had become the greatest evangelist the world has ever seen in a short two years. Now there was a big charismatic conference in Mexico City. And I was going with Ralph Wilkerson, David Wilkerson, Dennis Bennett, David du Plessis. I was on the team. I had it made.

It was the evening of February 9, 1972 and as it wore on, I became more and more uncomfortable, having some pain in my left wrist, and also in my chest. It was difficult for me to get a deep breath.

I wasn't doing a good job of masking my distress, and my wife, Betty, asked if she should call our doctor.

"Don't be silly," I told her. "I'm just having some gas pains." It was soon evident that it might be more than this, however, so I gave in and let her call Dr. Pete Wyatt.

He told me to come to the emergency room at Anaheim Memorial Hospital. Because it was late and all our children were in bed, I drove myself to the hospital. I approached the hospital with an air of unconcern; after all, the Lord was on my side. He knew how badly I would be needed in Mexico City. I was going to be the only completed Jew there. We are few and far between, and I knew the Lord was aware of it.

In spite of my protests that I was really quite all right, the doctor insisted on making sure. First he did a cardiogram, which was all right. Second, he had X-rays taken, which were all right. Third, he drew blood for a battery of tests. This didn't prove to be all right.

Dr. Wyatt explained to me that when a heart is in trouble, and a coronary is on its way, the heart releases enzymes, which show up in the blood. The blood tests indicated I was beginning to have a heart attack, and Pete wanted me in intensive care immediately.

My reaction was a flat no. I tried to explain that I was going to leave on Monday for Mexico City. I told him how indispensible I was to this meeting. After a great deal of fruitless arguing, Pete gave up in disgust. He finally said, "Mike, before you walk out of here tonight, I want you to sign a release absolving the hospital and me of any responsibility for what happens to you."

As I stood there, the pain was slowly building again, but I refused to give in to it. With the sure knowledge that I was the perfect overcomer, I scrawled my name across the bottom of the release.

On the drive back home, I began to plan on how I would handle Betty. First, I decided to take a codein tablet so I could fool her into thinking there was no more pain.

But Betty Esses was not a woman to be fooled. The next morning while Mike was installing drapery rods at Dr. Irvine Harrison's home, she called Dr. Wyatt.

"Betty," the doctor said ominously, "Mike is having a heart attack, right now. If he goes to Mexico City with its high altitude, you will probably have to go and bring his body back home."

It was a very nonchalant woman who called me at the Harrison's soon after my arrival.

Betty said, quite casually, "Mike, after you and Bob De-Blase get through fixing that rod, and you kibitz awhile, would you do me a favor?" She continued, "Pete wants to do one more blood test before you leave for Mexico. Will you guys drop by and have that done before you come home?"

Bob and I sauntered into the lab at the hospital, and I told them that Dr. Wyatt had ordered a blood test. I gave the girl my name; then we sat down and waited for her to draw the blood.

Suddenly, out of the blue, and in front of Bob's astonished eyes, I was shoved into a wheelchair by an orderly, wheeled round the corner to ICU, stripped of my clothes, put to bed, hooked up to the heart monitor, and an I.V. tube was placed in my arm. This was so they would have quick access to my vein if they needed it.

In the meantime, Betty was on her way to the hospital. When she arrived, she went to the office and signed me in, as pre-arranged. That was pretty slick. My only advice to any man who wants to fool his wife is — forget it.

When Betty walked into my room, I was lying there a veritable caldron of emotions. I was mad. I was scared. I was in pain. I couldn't understand why the Lord was permitting this to happen, but I was still determined to go to Mexico on Sunday.

All of this happened on Friday. That evening, Pastor Wilkerson came by. I was glad to see him, and have him pray for me. I told him several times that he was not to worry; I was going to be with him on the plane Monday.

Sunday morning came, and I was on needles and pins waiting for Pete to come in to release me. When he arrived, he said, "Mike, I want you here one more day. If necessary, you can go to the airport from the hospital." I didn't like it, but there wasn't much I could do about it, so I settled down to wait.

That afternoon, Grace Robley and Elizabeth Kitson came by. In ICU, only members of your immediate family or your minister are allowed to visit. The nurse made an exception to the rule when she was told that these women had come all the way from San Diego to see me.

Grace and her husband, Rob, who is a medical doctor, live in San Diego. They attended my classes at Melodyland School of the Bible.

"Mike, they only let us in for five minutes, so let me pray for you," Grace said. As Grace began to pray, a stillness came over the room, and I guess this is when the Holy Spirit began to talk to her.

Grace didn't say anything at all to me, but as soon as she left, she headed straight for a telephone to call Betty. "Betty, I have something to tell you. I have just been in to pray with Mike, and the Lord let me know that He is going to take Mike home. I have never known anything more clearly. The Lord is going to take Mike home, unless you can change His mind."

Totally bewildered, Betty asked, "How in the world could I do that?"

Grace told her to call every place where I had spoken, everyone she knew, to ask them to start praying for "the resurrection power" for Mike.

For the next two hours Betty was on the phone. She called every place she could remember me preaching. Sacramento, San Jose, Eugene, Pittsburgh, Chicago . . . She called Oral Roberts University. They told her they would put the prayer for resurrection power on a 24-hour prayer chain.

Many other churches said they would be praying round the clock. Even some of the religious radio and television stations were alerted.

Betty came by Sunday evening to see me, and I was still all right. I wondered why Betty seemed so nervous. Betty reluctantly left to go home, as I learned later, to spend the rest of the night on her knees praying for me.

That night, the roof caved in on me. I went so sour it was unbelievable. The pains in my chest started to be much more severe, and panic set in.

By the next day, events were crowding in on top of one another, and I was soon reduced to a quivering mass of humanity. I developed double pneumonia and blood poisoning. My veins collapsed, and the danger of clots in my arteries became imminent.

Dr. Wyatt was trying every antibiotic he could think of, but nothing was having any effect. My temperature was holding at around 106°, and I was hyperventilating. Days became one big nightmare blur. I found myself stark naked, lying on an ice mattress. They cool these things down to 65°. It's hard to believe, but as cold as that thing was, I lay there and sweated.

Now the Lord began to get through to me. I was beginning to see the light. I was being put through God's humbling process, having once again gotten too big for my britches.

Because I had diarrhea from the antibiotics, the nurses had to clean me and change me like an infant many times every day. Each time, I could hear the Lord say to me, "Will you take this humiliation and praise Me for it?" I told Him I would, and I did.

The days blended together. Thursday morning arrived, and it was as if I was coming back from another planet. I looked around me, and for the first time in days, I was aware of my surroundings. I began to follow the nurses around with my eyes as they accomplished their morning duties.

It was a funny sensation. I was aware of them, but they didn't seem to know it. I heard one of the nurses tell the other, "You better call the pastor's wife, because he's failing fast."

A short time later, Betty's face swam into view. I could

see that she was crying, but I didn't know why. I was feeling so much better.

She kept saying, "Fight, Mike, for God's sake, fight. I can't go on if you give up. You've got to come home for the kids and me."

In a few minutes, I looked up at one of the pastors from Melodyland. This Baptist minister leaned over the bed and clasped my hand in his. In a soft voice he said, "Can you hear me, Mike?" I tried to tell him that I could, and I think he heard me because he started talking very intensely.

He said, "Mike, I'm going to claim the resurrection power of Jesus Christ flowing through your body. Do you understand me, Mike?" I nodded yes.

He said, "Do you understand that I'm going to bind Satan and the friends of Job who are telling you to give up?"

I nodded yes again.

Then the pastor began to pray the prayer that God had given him to pray. As he prayed, I became more and more aware of what had been happening. I hadn't been coming back to life. I had been drifting into death, on a cloud of lethargy. I hadn't been fighting at all.

As I watched the pastor walk out the door, panic inundated me. I began to hyperventilate again.

One of the nurses ran to the bed, and with a voice of authority, she said, "Pastor, in the name of Him you have been telling me about, I command you to stop this hyperventilation."

I was so surprised to hear her use these words that I must have just stared at her. She said it again with even more authority: "Pastor, in the name of Him you've been telling me about, I command you to stop this hyperventilating."

I stopped, and the awareness came that my work was not done for the Lord.

I cannot say whether Dr. Esses was clinically dead the afternoon I walked into his ward. But the death certificate was there on the table. I grabbed his hand and started thanking the Lord, remembering God had extended him before; God had a plan for his life and the Lord was not through with him. Then I began to quote Bible verses.

I am the resurrection, and the life: he that believeth in me, though he were dead, yet shall he live . . . Behold, I give unto you power . . . over all the power of the enemy . . . He that believeth on me, the works that I do shall he do also; and greater works than these shall he do; because I go unto my Father. And whatsoever ye shall ask in my name, that will I do, that the Father may be glorified in the Son . . . where two or three are gathered together in my name, there am I in the midst of them.[3]

I was oblivious to anybody around me. I was sharing with God in that room.

Mike opened his eyes and smiled at me when I finished praying. "Thank you, Pastor," he said.

I knew it was done. It wouldn't make any difference what the doctor said, I knew it was done. I could see the color coming back into his face.

In a few days his symptoms left. Slowly, his temperature dropped. His heart stabilized, and the double pneumonia was gone. His veins were normal again, and there were no signs of blood clots.

After 33 days, 11 in Intensive Care and 22 in a regular ward, he went home from the hospital.

Before his experience, Mike had an aggressive, driving personality. He manipulated people to accomplish his purpose. But his life is transformed now. He's gentle. There's a softness in his approach. He doesn't manipulate people anymore.

Dr. Esses still has some of the same problems he had before. He still has a handicapped child. He still has deadlines to meet, but he has learned to turn his pressures over to God. Before, he would worry. Now, he has peace and joy in his circumstances.

One cannot experience resurrection without a dramatic change in his life. In the next chapter I will discuss eight significant changes, which I've noted in those who have had beyond and back experiences.

4
"I Hovered Over My Body"

"Do you get a heartbeat?"

"Not a sign."

"We have tried so hard to save them."

"But we've lost them both."

This dialogue of death continued between the doctors as they anxiously labored to save a mother and child at the time of delivery.

Let me tell you how I heard this story.

We were visiting our friends, Jim and Kay Lambeth in North Carolina. He is a legislator of that state. And since the legislature was in session, we met them at their hotel. The entire seventh floor of the Raleigh Hilton was occupied by the state lawmakers. They jokingly call this floor "Seventh Heaven."

In the Lambeth's room, waiting to go to a reception, I mentioned to Kay that this book was in the early stages of writing.

"Ralph, all my friends are talking about out-of-the-body experiences. Have I told you mine?"

She started to relate her intriguing story.

"Just a minute," I interrupted. "Let me snap on the tape recorder."

Said Kay Lambeth:

It was what I call astral projection because I have no other name for it. Although it seems as if it were yesterday, I had this experience 25 years ago.

I was pregnant with my last boy and was at home when I hemorrhaged. My brother, who was a medical doctor, came immediately when I called him.

"Well, you have to go to the hospital," he said authoritatively. That was such a shock to me, but I was losing so much blood he was not at all sure what was going to happen.

"It doesn't look like we're going to be able to save the baby," he told me. "But we'll do everything we can to save you."

"Please save the baby," I pleaded — that was my whole thought. I didn't think about being afraid. I didn't think about anything except, "God save the child. This child deserves an opportunity to live. And what a joy he'll be."

He took me to the hospital and got in touch with a surgeon who examined me. By that time, of course, my blood count was apparently very low, and they didn't know what they could do.

They decided to do a Caesarean section on me, and took me into the operating room to put me to sleep. But whatever they gave me did not react.

As far as I am concerned when they started talking over me, back and forth with the nurses, I still was not asleep. I couldn't go to sleep, and I realized they were about to start operating on me.

At that time I left my body and stood at the foot of the operating table with the doctor. I not only stood there, but felt the pain from the body while I observed it. I felt that knife cut me.

The pain was excruciating, and there was nothing I could do but be there. Then I began to realize that I was not in that body, but all that pain was going on. I tried desperately to get word to the doctor. I wanted to tell him so badly that I could feel the pain.

I remember they took the baby; it was not alive. They put him on the table, and a doctor started breathing into him. Then I heard him cry. I heard the doctor say, "Nurse, put the baby in the incubator. I must tend to the mother. It doesn't look like she's going to make it."

I thought, isn't that strange? All that pain she's going through. I was having the pain for the person, but I was not in her body. I watched the whole proceedings. Soon they took her back to the room.

I had to tell my brother because I knew that I was not going to be able to stay on this Earth. I knew there was no way to get back there, but because my brother was a doctor, I wanted to tell him so he'd know how to tell other people to watch what they did, and not take things lightly. You must completely fill your minds with good things. So I tried so desperately to get this through to him.

He kept saying, "It's all right, it's all right, whatever you've taken."

I said, "Herman, I'm not going to be able to stay back on this Earth, but I want you to know these things. I have to tell you now because there's a voice that keeps saying, 'You know, now you've been there, but you will not be able to remember it to tell anybody else.' " But I was determined to tell him.

I can't remember certain areas that happened to me. But I do recall there was a great deal of light. Beautiful light. But the voice was contrary to the light.

When I realized that I was coming to, I was frantic. I came back into my own body the next day. They kept saying to me, "You're all right."

I asked about the baby, and they said they didn't know if he was going to live. They had put him in an incubator, but he wasn't doing well.

In any event, we all got well. My experience never left me. It was so real, I knew I had left the body. I know what happened. I saw the light.

I don't know whether the records would say that I was clinically dead. But my brother knew. He said that at one time I was gone. I presume I was because they were working frantically on me with adrenalin and oxygen to bring me back to life.

Although this happened 25 years ago, every time I've been to the hospital, I've asked the doctors, "Can you give me something so that I'll be sure that I don't get to that point again. I don't want to leave my body and observe this. I want to be asleep or I want to be gone to Heaven. I don't want that in-between state." No one's ever been able to help me. They just say, "Oh, well, that sometimes happens when you take Pentothal." But that is not true of what happened to me. I left my body.

Multiple out-of-the-body experiences are being told today. A fine line of distinction exists between the occult and the spiritual. Even as the magicians of Moses' day duplicated supernatural acts, Satanic power today can still produce the counterfeit.

Out-of-the-body experiences reported both by Christians and those caught up in the occult have striking similarities. Here are some:

1. They float over their body.

2. They travel through a dark tunnel.

3. They see a bright light.

4. They hear buzzing noises.

5. They see departed loved ones.

But what are the distinctive differences?

1. Both have the sense of being caught up, viewing things still going on here in this world. Usually, neither wants to come back. The Christian is sublimely happy with his Lord. But the non-Christian's bliss is delusion. The happiness cult fits the Bible verse that says, "There is a way that seemeth right unto a man, but the end thereof are the ways of death."[1]

2. David, in the 23rd Psalm, described a dark tunnel when he mentioned walking through "the valley of the shadow of death." But he said he would have no fear because the Lord was with him. The non-Christian walks through the same tunnel, but ends up in an illusionary place.

3. Do not be deceived by lights. The Bible says Satan can transform himself into an angel of light. Jesus declared, "I am the light of the world." Spiritual discernment is absolutely necessary to distinguish between the two.

4. A Christian often hears the sound of praise. St. John in the Book of Revelation said he heard a voice "as of many waters" praising God. The buzzing reported by occultists is a sound of confusion producing fear, not faith.

5. Both see departed ones. In spiritualism people claim they call back the dead. This is an illusion. Through a medium Saul thought he had called back Samuel, but learned later that it was an evil spirit. The Bible strongly forbids consulting with fortune tellers, astrologers and seers (Deuteronomy 18:12; II Chronicles 33:6). Could all the loved ones seen by non-Christians be an evil impersonation? Departed loved ones seen by Christians are not disembodied spirits, but real persons at home with the Lord.

The final difference between these contrasting descriptions of destiny is that God awaits the arrival of the Christian in Heaven. The non-Christian has an abstract and unreal image of the beyond.

Kay Lambeth went on to say she now realized her major fears were produced by an occultic out-of-the-body experience. Since her commitment to Christ, she knows her destiny and is no longer haunted by this possibility.

Psychological and Spiritual Changes

A fascinating and inspiring pattern of personality changes has emerged from the experiences of Christians who have been beyond and back.

Many have testified to losing their fear of death. Some, like Marvin Ford, have become more practical and life-related in their day-to-day activities. Others acknowledge losing racial prejudice and finding a new love for all people. A greater ability to understand God's Word has been reported by others.

At least nine basic psychological and spiritual patterns have emerged from these case histories: life and family in proper focus, materialism no longer an obsession, a greater sensitivity to the Holy Spirit, a spirit of revelation, a global vision, no fear of death, a more mellow nature and a new sense of joy.

Let's examine each of these changes.

Life in Proper Focus

When a person comes back from the dead, he has a new sense of values. The dream house with the fireplace and light blue carpeting, sleek new automobile parked in the driveway, twin-engine fiberglass cruiser bobbing in the marina—no longer are these a priority.

Usually the first change in values is a person's schedule. God moves into first place, family becomes second, the church third, and employment last.

Some tell me they are taking the time now to do things they had always intended to do, but never found the time. Dr. Irvine Harrison was one. Immediately after his return from the beyond, he and his wife flew to South America.

After Dr. Harrison's resurrection from the dead, they visit with natives of South America during a missionary trip.

Though he was a world traveler, he had never taken time for
a leisurely trip. While this was vacation, it was vacation with
a purpose, for he ministered to many missionaries and friends.
People often put off a vacation or a trip because they think
they don't have the money or the time. My advice is don't
put off living until you think you're going to die — or until
you die. *Live now*.

A woman told my wife, "I'm taking a vacation. I really
can't afford it. But, it's cheaper than a doctor bill!"

A college-age friend of ours worked on a Northwest Passage cruise ship making runs to Alaska. This friend told us
that often one or two people die on a voyage. Most of these
people were elderly and had waited a whole lifetime, saving
their money for such an event — only to die on the trip.

It isn't selfish or wasteful to plan vacations while you are
younger and in good health. People think nothing of taking
out a loan to buy a boat, a cabin in the mountains, a vacation vehicle, build a room addition or install a patio. Yet
many would not think of borrowing money for a vacation.
Taking a holiday is practical, besides being enjoyable.

Dr. Christian Barnard and other noted surgeons were
making medical history with heart transplants about the time
Dr. Harrison died. One of the celebrated transplant cases was
Philip Blasberg, who had undergone the surgery on December 3, 1967 in Groote Schuur Hospital in Capetown, South
Africa. According to news stories, Blasberg had a personality
change after his transplant. For five and a half years, he lived
in misery, having moods of depression. At times he was like
a stranger to his family.

In contrast, Dr. Harrison lived for eight and a half years
after his resurrection in perfect health without being depressed or moody. His post-mortem life was full of excitement, travel and warm fellowship with friends.

Marvin Ford, too, had a new sense of values after his
back-to-life experience. A very active churchman, he had

served faithfully. After his beyond and back experience, he considered his work a ministry — not just a responsibility.

The changes I observed in Dr. Mike Esses were the most dramatic of any I've mentioned so far. He had a tendency to spin out things too fast. He was highly impatient. But since his experience, he has reassessed his priorities. Now Mike realizes he was trying to compress too much in too little time.

Most of us are plagued with busyness. Jesus had a world to win, but I never get the feeling that He was in a hurry. He had time for children, for fellowship, fishing, weddings — even vacations. Now Mike projects a similar relaxed attitude. I can sit down and talk to him. I couldn't do that before.

Family Comes Into Focus

Family is second only to God when our priorities are in order.

It's just as important to take time for your children's emotional and social needs as it is their material necessities. More important than providing a roof over their heads is taking time with the family.

Often I have stood by the grave and heard the lament of a loved one, "If I had only taken more time . . ." One of the greatest compliments my oldest daughter, Angela, paid to me was, "Daddy, even though you were a very busy pastor, you consistently made time for us. Some of my best memories are the annual fun vacations our family took together."

Mrs. Esses says love also is spelled t-i-m-e. For years Betty's complaint had been that her husband was seldom home. He didn't know his children. But that's not true now. Mike has cut back on his speaking engagements and is spending time at home with his family.

Dr. Harrison took more time with his sons' business projects after his resurrection. Before leading his last tour to the Holy Land, he and his son, Bob, finalized some business for Bob's new car leasing agency.

Taking time is an important principle here. Had Dr. Harrison decided to help Bob when he got back from the trip, the decisions they made might never have been resolved.

Some parents have family-centered interest, but they put off taking the time to develop them. Three four-letter words in family life are *time, love* and *care.*

Because of their musical interests the Fords were always close. But after his experience, he seemed to take more interest in relatives.

Materialism No Longer an Obsession

It's easy to get caught up in materialism. Advertisements in newspapers and magazines, on radio, television and billboards create an incessant desire for things. Newer cars, bigger homes, the latest time-saving appliances, white sales and back-to-school sales . . . the list is endless. Consequently, most people are trapped into long working hours, borrowing and juggling accounts to feed an insatiable hunger for material goods.

But with those who have had a beyond and back experience, materialism is no longer an obsession. Although Dr. Harrison was anxious to provide for his family, he was never obsessed with money. But security for his family was a strong drive. After his out-of-the-body experience, tension and worry over finances no longer pressed him.

Mike Esses was constantly striving in his business. He was both successful and aggressive. Trying to corner the market in his field, he worked long hours. But after God raised him up, Mike put a "For Sale" sign on the door of his business. He said to me, "Pastor, that's not really my thing now. My greatest desire is to proclaim Jesus Christ." Not that Mike didn't need the money . . . but being in full-time ministry, he now looked to God for all his needs.

Though materialism isn't a focus, inevitably when we put Christ first, God gives us back much more than we've had

before. Jesus said, "Seek first His kingdom and His righteousness; and all these things shall be added to you."[2]

Greater Sensitivity to the Holy Spirit

There's a difference in being filled with the Spirit and
flowing in the Spirit. Many don't comprehend this.

The concept of flowing in the Spirit was the greatest
revelation that Dr. Harrison had when he came back from
the dead.

"Ralph, God gave me a whole new insight. While I have
taught theology for many years, while I've been a Spirit-filled
man and walked in the Spirit for many years, while I have
been the president of several schools, I still did not understand what it was to move in this new dimension.

"But, Ralph, the day will come when we will stand before
tribes of people who have never heard the name of Jesus and
don't understand what we say. But the Holy Spirit will get
through to them. God will give us love and compassion, and
it will flow from us to them. Entire tribes will be converted."

Before his experience, he was a gifted, methodical theologian. After his resurrection, however, he began to see the
concept of ministering spirit to spirit.

This became the underlying philosophy of the Melodyland
School of Theology. We're not just talking about what some
famous historian or theologian said a hundred years ago.
We're concerned with what God's Holy Spirit is saying to
human needs today.

In each case where I stood at the bedside of a deceased
person and spoke Bible verses on resurrection, I did not speak
to his mind or intellect. I spoke to his spirit. Often I have
walked into a sick room and prayed while the patient was
either unconscious or sleeping. I found at times the results
were even greater than if he were awake or alert.

"But what is the *flow* of the Holy Spirit?" you ask.

It is not emotionalism. It is not coining a few words or

learning 10 easy steps in how to lead a person to Christ.

It *is* that indefinable quality of the uncreated energy of God moving through a human being to accomplish God's work. Many mature ministers, even those who have been deeply involved in spiritual life, do not comprehend this dimension. God's greatest work in the church is being done in this realm.

The word "spirit" in Greek is *pneuma,* from which we get the term *pneumatic* meaning "air." Just as one cannot see or touch air, so the flow of the Holy Spirit is intangible. But as one can feel the movement of air, he also can sense the flow of the wind of God.[3]

A Presbyterian minister relates how his church has come alive to new truths that he has preached for many years. People in a Catholic prayer group say that the sacraments have taken on a new meaning. Businessmen report that the Sunday sermons are now being translated into everyday business.

Leaders from around the world have visited Melodyland to see the church in action. Their first questions usually are, "Do you use computers? How many buses do you operate? What is your stewardship program? How large is your budget? How many people do you have on staff? How large is your Sunday School?"

To these questions I respond, "None of these has built this church."

Recently a newspaper reporter asked me, "To what *do* you attribute the growth of Melodyland?"

I answered, "We exalt Jesus in every service."

"So do other churches," he challenged.

"But we are a Bible-believing church," I returned.

"Most churches claim that, too," he persisted. "Maybe it's your personal charisma or leadership qualities."

"I deny that, too."

"Well, tell me! What is it?"

"It's the conscious, unseen presence of the Holy Spirit that the people sense in the services. The flow of the Spirit meeting human need is the great supernatural attraction that draws people back Sunday after Sunday. This same spirit could and should flow in every church around the world."

It isn't *what* we say that convinces others, but it is the Holy Spirit who communicates. Marvin Ford is an example of this. He ministered at Melodyland one Sunday evening recently. One of the individuals who received a Christian healing that night commented, "It wasn't so much *what* he said, but *how* he said it." It was the Spirit behind what Marvin said that moved people.

Mike Esses also has a greater sensitivity to the Spirit. He can weep unashamedly now. Weeping men are winning men. Souls are saved at the cost of tears. Such sensitivity has helped men become better communicators.

In his last sermon at Melodyland Dr. Harrison taught this new means of ministry. When he talked about Jesus, you visualized our Lord Jesus standing beside him. He introduced Jesus, not just theology.

Spirit of Revelation

The spirit of revelation is one of the gifts of the Holy Spirit called the *word of knowledge*.[4] I prefer to call it a *spirit of knowing* because you know things about people that aren't revealed except by the Spirit.

Insight into this began when I was in the hospital a few years ago to undergo two operations. During this time, I prayed day and night, and the Lord began to talk with me. He let me see people and look beyond the external personality.

Many times smiling people actually might be crying inside. Christ's compassion gave me fuller understanding of the inner man.

Unable to sleep one night. I heard the patient in the next bed stirring as though in pain. We had a brief conversation when he was moved into the room. Though he told me he was terminally ill with cancer, we did not discuss either of our backgrounds. A supernatural spirit of revelation flashed before me. "Sir, you are the son of a Russian pastor from the old country, and you are running from God. But God is answering your father's prayers."

"Who told you that?" His voice broke. "I've never even told my wife this. She knows nothing about my family still in Russia."

"I don't know anything about your past, but God does," I replied. "And I'll tell you something else, you might have given up God, but He hasn't given you up."

The man prayed the sinner's prayer with me. A few hours later he died.

A classic biblical example is the woman at the well. Through the spirit of revelation, Jesus shocked her by revealing her past life. She had had five husbands, and the man she was living with was not one of them. Our Lord knew everything about those to whom He ministered.

Today this same gift must operate in us. Some object, "But we are not Jesus!" True, but we are God's people. Jesus declares, "Even as the Father has sent me, so send I you."

People coming in for counsel often will not disclose the real problem. But a spirit of knowing can identify the deeper need.

I remember one day when Dr. Harrison was counseling with a person at Melodyland. He was getting nowhere because the person would not tell the truth. By the spirit of knowing, Dr. Harrison identified the problem. As a result, that person broke and there was a healing.

Dr. Esses, too, has this sensitive ability now to perceive a person's need and minister to him. God revealed to Marvin Ford the exact needs of some people in the audience at Melodyland. Except by a gift of the Holy Spirit, he could not have known specific details.

Don't confuse this sacred gift with psychic predictions of air crashes, earthquakes, assassinations and personal fate. These prophecies are Satanically inspired. The true gift of the spirit of knowing is a divine impartation of God. It is not obtained by natural wisdom, speculation or suspicion. Neither are you born with it, nor can you cultivate it.

All gifts of the Spirit are supernatural. *Supernatural* means just more natural. Super natural. Too often we think the gifts only operate inside a church building. But when needed, they are available in everyday life.

Global Vision

I recall Marvin telling me of his new commission to reach the world. I blinked in surprise, thinking, "You've never been an evangelist; you've never been away to school. How could *you* reach the world?"

Suddenly, I was aware that all Marvin has to do is tell people what happened to him. God would do the rest.

I've already told Dr. Harrison's vision of reaching primitive tribes. And Dr. Esses' itinerary takes him to many parts of the world, teaching the Bible.

Melodyland is committed to reach this generation. Its witness is global.

No Fear of Death

Once a person has experienced death and returned, it is no longer an unknown. All fear stems from the fear of death, as we shall see in Chapter 12.

A loss of the fear of death is a change frequently reported by those who have had out-of-the-body experiences. Sometimes this change surfaces in related fears such as flying. Mike would not face this phobia until some time after his out-of-the-body experience.

Every time he had a scheduled trip, he would drum up excuses to keep from going. He would even drive a thousand miles to avoid flying. The only way his wife, Betty, could get him on a plane was to fill him with Dramamine.

When we got on a plane, cold sweat would bead on his face. He would grip the arm rest on the seat until his knuckles went white.

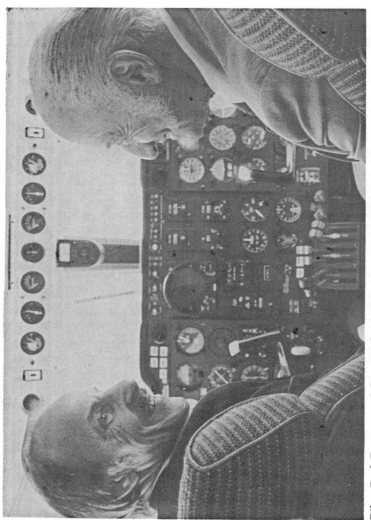

Pilot Bud Darvas and Dr. Michael Esses in one of Melodyland's planes. Dr. Esses no longer fears flying.

It was after his encounter with death while we were flying back from Pittsburgh, Pennsylvania, that I said to him, "Mike, you've already faced death and God brought you through. The life you are now living is a gift of God. It's not God's plan for you, a man of faith, to live in fear. You're discounting the whole message. God doesn't intend for any Christian to live under the spirit of fear."

I looked at him intently for a moment, then continued, "Mike, if you are God's man, with God's message, God has an insurance policy on you, and nothing can happen to you until it's His time. Of course, if you do anything foolish like flying with an inexperienced pilot or taking chances in marginal weather or with bad equipment, you're tempting God."

I took hold of his hand. "Mike, let's agree that you'll never have the fear of flying again."

Within five minutes after we prayed, color had come into his face. That ashen look of fear was gone, and he had relaxed his grip on the arm rest.

Glancing at a family sitting across the aisle, Mike saw little children who reminded him of his family. To my amazement, within 30 minutes he was playing in the aisle with the children, totally unconcerned that we were at 39,000 feet.

More Mellow Nature

Another observation of each of these men is that their natures became gentler. They mellowed. In taming us God harnesses our willful, rebellious ways. Dr. Esses, for example, manipulated people for his own good, but now he's more considerate of others. Dr. Harrison was always a poised gentleman, but after his experience he no longer struggled against time. There was even more quietness in his spirit.

Approaching the end of the journey of life, when you're facing the realities of the next world, your disposition becomes more in accord with the nature of Christ.

New Sense of Joy

Joy is a characteristic of life in the Spirit. The world seeks true joy and happiness more than any other pleasure.

The early church was known for its joy. Communicating joy and happiness is still our greatest advertisement for reaching the world. Joy does not depend on circumstances. Paul and Silas in the Philippian jail had every circumstance against them, yet they were joyful and sang praises at the midnight hour.

How could they have such joy? Because they had a sense of destiny and were certain that God had everything under control.

Happiness depends on happenings. Every year at Christmas time the Harrisons hosted a party. It was a time of celebration. Favorite food—salads, ham, turkey and all the trimmings—were in abundance. More than 200 different friends came each year, and the Harrisons would say, "This is our gift to you."

The purpose of the party was to let their friends see the radiance of Christ. A happy family is one of the most tremendous testimonies in a confused world. I'm certain that a number of people will meet Dr. Harrison in Heaven because of his joyful witness.

Both Marvin and Mike exhibit the same exuberance.

The joyous Christian doesn't have fewer problems than the non-Christians, but he doesn't permit circumstances to rob him of his joy.

Coming up in the next few pages will be some almost incredible incidents that happened in other countries. None of these miracles took place at church, but God gave ordinary people extraordinary power—even to raise the dead. Intuitively, most assume that miracles are reserved for the minister. Not true! God can use *you.*

5
"I Saw the Living Dead"

People being raised from the dead in Indonesia! Spectacular! Sensational! It sounded too good to be true.

I began asking God if I should go to investigate for myself. As a student of revival, I wanted to do more than just read about the great spiritual awakening in Indonesia. I wanted to see it. I wanted to learn some of the principles of that renewal.

As Mrs. Wilkerson and I prayed, it became clear to us that God was saying, "Go!"

Target Timor

A remote island 750 miles from Java was the geographical heart of that phenomenal revival. So we set out on our first trip to Indonesia in 1968. Exactly how we were going to get to tiny Timor we didn't know.

Flying first to Djakarta, we asked the travel agent about air transportation. His look suggested that we were out of our minds. "Why would you want to go way out *there?*"

"Because I've read about the great revival that's going on. The dead are being raised. I want to interview some of the living dead."

He shrugged his shoulders. "There's no direct flight there. You'll have to fly to Bali first, and the plane only goes when it has a full load. There are no public hotels in Kupang where the only airport is. I don't even know where you'll sleep. You realize, of course, that this island is still very primitive. I do know a government official who lives on the island. Here's his name. Maybe he can help you."

On the first leg of our trip we stopped in Bali, an exotic resort island. With nothing to do but wait, one day I picked up the phone book and began leafing through it. I know the Lord helped me to figure out the word *church* in Indonesian. I called the number of a church at random.

A miracle! A man answered, "Hello!" in English. He was a Christian Missionary Alliance pastor.

"Do you know anybody on the island of Timor?" I asked after identifying myself.

"No, but I'll check around."

I invited him and his wife to dinner. At the table I asked, "What do you know about the revival on Timor?"

"Revival? On Timor?" He looked puzzled. "I don't know what you're talking about. But that's not strange. You can't imagine how bad communications are out here. Same with transportation. But I guess you've already learned about that," he grinned.

I said, "We've come all the way from Los Angeles to find out about this revival. We must get to Timor, for we know God is sending us."

Another miracle! The missionary had learned the name of an English teacher on the island who could be our guide and interpreter. At last we boarded an old C-47 cargo plane for the flight to Timor. The plane had cot-like canvas seats and made stops at five small islands.

Again, another miracle! As we stepped into a tiny building that was the terminal at Kupang, a distinguished gentleman said, "Greetings, Pastor and Mrs. Wilkerson. My friend, the travel agent in Djakarta, got word to me that you'd be coming in today."

To our surprise, we learned that he had been involved in the revival and could help us with details. He spoke perfect English, and enabled us to locate Paul, our interpreter.

When I told Paul why we came to Timor, he hung his head. "God called me to be part of this revival," he lamented. "But I got too busy."

Paul helped us look for a room. They told us we could stay at one place, but it was filthy! An open latrine was right in the house. The sleeping quarters were common rooms with only curtains separating the beds. Chickens and pigs wandered in and out.

I arranged for a jeep to take us on to Soe. It was a leftover piece of junk abandoned by American G.I.s years before, but it was the only vehicle for hire. I also heard the road was just a trail over very rugged terrain. Now I had a new challenge. I couldn't possibly subject my wife, Allene, to the rough trip. But where could I leave her?

We went back to the man who had greeted us at the airport.

"I'm building a new house," he offered. "The roof is on and the flooring is in, but it's not finished. I keep a night watchman, so your wife could stay there safely while you're away."

We gratefully accepted. Then came the biggest surprise. He moved in a brand new bedroom set purchased for their new home. God had provided more than just shelter—a new, immaculate mattress with *no bugs!*

Everyplace we went people turned and stared. I thought it was because we were Americans. But it wasn't so. It was Allene. One interpreter explained that many of the villagers had not seen a blue-eyed, blonde-haired person before. They'd stop, point and chatter among themselves.

The main trail went right by our bedroom, so a night watchman stood guard.

I hated to leave my wife, but there was nothing else I could do. I committed her to the Lord and set out at four a.m. the next day by jeep for Soe.

The trip was 60 miles. Sometimes there wasn't even a trail; we had to make our own way and ford rushing streams. The jeep broke down at least 30 times. Thoroughly exhausted, we arrived in Soe at three o'clock.

God was ahead of me. He had given a vision to one of the Christians, telling him that I was coming. The villager told his pastor, and 30 persons had gathered at the pastor's house to meet me. Even the villagers whom I wanted to interview were there.

The pastor's house was meager, but the islanders considered it the nicest in the village—it had running water. When a lever was pulled, the water ran down a trough from a storage tank on the roof.

"God has sent you to us . . ." the pastor greeted, gripping my hand warmly. "We have this report, the one about all the things God is doing on Timor. Brother Daniels, our superintendent, was supposed to have taken it to Singapore for the meeting."

"You mean the Billy Graham Congress on Evangelism?" I asked. (I was invited to that historic meeting, so I knew what he was talking about.)

"Yes," he nodded, "but because of the political situation, Brother Daniels couldn't get out of the country." Tears began to stream down his face as he handed me the report written in

Indonesian. "Get this to the world; it *must* get out to the world!" he pleaded.

That day, I talked personally with some who had witnessed the miracles. The two children who had been raised from the dead were among those I interviewed through the interpreter. Mother Sarlin, who prayed over the children, was there, too. When I asked her about the resurrections, she hung her head, confessing that she had started charging money to pray for miracles and God had not used her since.

Can you imagine my thrill that afternoon as I began to hear them relate exciting miracles paralleling the New Testament—walking on water, turning water into wine, light by night, resurrections, being supernaturally transported from one place to another, and spectacular healings?

Talking softly, they would look at the floor, avoiding any eye contact. I sensed their hesitation in repeating the miracles, for fear of taking the glory. That impressed me the most. My particular interest in those raised from the dead was surprising to them. They didn't consider that any greater than the other miracles.

In that meeting were four of the eight resurrected persons I met on the island. Even *they* didn't project a greater sense of awe when talking about that miracle. Their attitudes convinced me these reports are true.

It was nearly dark when we reached the church. I expected to see a handful of people, but the church was crowded with more than 400 praying worshipers.

"Why are they here? It isn't a regular service," I said.

My guide smiled knowingly, "They are praying for the teams."

"The teams are still going out?"

"Yes. They go out early every morning all over the island to tell people about Jesus. These people back them up with prayer."

I will never forget praying with those humble, Bible believing people. Laying their hands upon me, they prayed quietly, but powerfully, in their own language. Somehow I knew what they were praying!

Later they took me to see their Bible school, which was a crude, unfinished building with dirt floors and rock walls. But they were so proud of it.

"When will the roof go on?" I asked.

"As soon as God provides the money," the pastor smiled. "You see, we must have a tin roof because of heavy rains and strong winds. . . ."

"How much will the roof cost?"

"Much money, Pastor. About a thousand dollars."

I reached into my pocket and pulled out a check. Before I left, our small church had given me $1,000 to "use however God directs." Signing it, I handed it to the pastor. To them it was a great miracle that God had sent me to supply their need, for it would be the monsoon season shortly.

On the long, bumpy ride back that night, I was still talking about the miraculous.

"Yes, it is great," agreed a young pastor who made the trip with us. "But there are greater miracles. . . ."

"What are they?"

He looked puzzled that I should ask. "Why, it's when people come to Jesus . . . when their lives are changed . . . when they burn their fetishes and idols!"

I understood . . . and agreed . . .

Supernatural happenings first appeared among Indonesian believers when one Timorese pastor, J. M. E. Daniel, head of the presbytery in Soe, prayed fervently with his congregation that God would perform miracles among them.

The Lord appeared in a vision to a pastor's son, a young teacher named Ratuwalu in 1964, promising him a healing ministry on Timor. About the same time on Java, God began to work among the teachers and students of a Bible Institute.

The following summer, students in teams went throughout

the islands proclaiming the Gospel. Soon 150 teams were at work on Timor. Within two years, 150,000 persons found Christ, including the king.

Documented Miracles

As I mentioned earlier, many of the miracles witnessed by the teams were documented in the report prepared for the Billy Graham Congress in Singapore.

Let's examine some of these team reports, which were translated from the document by the Missionary in Bali.

Water Into Wine

The turning of water into wine is one of the most talked about miracles of the awakening.

Grapes don't grow in Indonesia, and watered-down imported liquor was used in Communion services.

The miracle first occurred in October 1967.

The Lord revealed Himself in a vision to three believers, Terah Selan, Johanna Nomleni and S. Boimao. In an audible voice He told them to wrestle in prayer in order to perform this miracle.

"You must wash yourselves and cleanse your hearts by prayer, confessing sin," He told them. "And you must be ready to go where I send you to get the water."

The Lord showed them a certain spring some distance from the house where they met to pray. At midnight, after much intercession, they were directed to take a water vessel and go to the spring.

There they knelt and prayed before filling the vessel. On the way home, they were led by a supernatural light. Occasionally, they said, "We saw the Lord Jesus walking in front of us with a stick in His hand."

Back at the house again they prayed over the water, sang hymns and read the Bible for 36 hours. Then the Lord spoke again.

"Look at the water."

It was wine.[1]

Just a few weeks ago, Nahor, a former school teacher who was the first one used in raising the dead, wrote to me. He mentioned that they are still seeing water being turned into wine.

Light by Night

Team 31, led by 33-year old Zacharias Nebuasa, was guided by a ray of light by night to a certain village to preach.

The team was caught in a thunderstorm. Though members had no umbrellas, they didn't get wet.

Several people from a church in Soe did not believe in the reports of healings and resurrections. Among the dissenters was a village elder who had been instrumental in building the church in Soe. He was angered one evening after listening to team testimonies in the church.

As he walked along the road, a poisonous snake bit his foot, and immediately it began to swell. During the week, his condition became worse and the swelling went to his arm. Unable to walk, he was carried on a stretcher to church the following Sunday.

He asked the team to pray for him. Following the leading of the Holy Spirit, they told him he had been bitten because of his unbelief.

This he acknowledged, and he was healed immediately.

Another young man from the same church opposed the team's witness and made fun of them. He still used charms and amulets and did not want to give them up.

At home after the service, he was stricken with a serious stomach disorder and nearly died. His parents quickly called the team to pray and he was healed.

After repenting of his sin, he surrendered his charms to be burned.

On the island of Sumba one night members of a Gospel

team were returning to their village. It was about one o'clock in the morning.

They approached a river they had crossed earlier by dugout canoe, only to find that the boat had been taken to the other side for the night.

They stood there for a moment wondering how they could cross. Spontaneously they broke into singing, and worshiping God. Suddenly they started walking on the water and reached the other side without sinking.[2]

Supernatural Transport

On another occasion a team led by a young woman named Judith Pobas was threatened by unbelievers who met one night to plot against the team. As these people met in an ex-army corporal's house, planning legal action against the Christians, the Lord impressed Judith and the team to pray.

As they prayed, the Holy Spirit snatched Judith away[3] and took her to the center of the meeting. She could see and hear everything, although she was not seen herself.

"Pray that they will get sleepy," the Lord directed.

After Judith prayed, the men began to nod, and their discussion faded. Suddenly the corporal said, "Why are we getting so excited about these crazy people? Let's not bother."

In similar incidents the Lord took the team leader and used her prayer to foil the plans of the enemies.

Resurrections

Resurrections from the dead also were reported. A four-year-old boy had been dead for four days. He was laid out on a slab while villagers mourned. When Mother Sarlin was one of the team leaders, she felt led of God to go and pray for the child. As she did, he suddenly came to life.

"All of the people knew that the boy was dead," says missionary Robert Little.

On another occasion, a little girl, whose parents were

leaders of Team 38, went into unconsciousness for two days. Ants crawled about her eyes. Again, Mother Sarlin was called to pray. The child was restored to life.[4]

Miss Susanna Nubatonis, 25 years old and unable to read or write, was the leader of Team 48. She prayed for a four-month-old baby who had died, and the infant came back to life.

Mel Tari, a young man who was a member of another team, gives this back-to-life account in his book *Like a Mighty Wind.*

> We were in a village in the Amfoang district where a man had died. He had been dead for two days. The family invited us to the funeral because there were many people planning to come, and they said, "Maybe you would have a word of comfort to give to the family."

> When we arrived there, there were more than a thousand people. That man had been dead for two days and was very stinky. In our tropical country, when you're dead six hours, you start to decay. You couldn't stand within 100 feet of him. The people there just look terrible in two days after they have died.

> When we were there and sitting with the mourners, suddenly the Lord said, "Please stand around that dead person, sing songs, and I will raise him back from the dead."

> I said to the Lord, "Oh, Lord, please give me a simple heart, and move in our midst."

> We went and stood around this dead person. We began to sing. But after the first song, nothing had taken place. So we started to wonder, "Lord, if You're going to raise him up, please do it quickly because we can't stand to stay around this stinking man. . . ."

> Then we sang a second song, and nothing happened. On the fifth song, nothing happened. But on the sixth song, that

man began to move his toes—and the team began to get scared.

We have a story in Indonesia that sometimes when people die they wake up and hug a person by their coffin and then die again. However, we just went ahead and sang. When we sang the seventh and eighth songs, that brother woke up, looked around and smiled.

He didn't hug anybody. He just opened his mouth and said, "Jesus has brought me back to life! Brothers and sisters, I want to tell you something. First, life never ends when you die. I've been dead for two days and I've experienced it."

The second thing he said was, "Hell and Heaven are real. I have experienced it. If you don't find Jesus in this life, you will never go to Heaven. You will be condemned to Hell for sure."

After he had said these things, we opened our Bibles and confirmed his testimony by the Word of God. In the next three months, as our teams ministered in that area, more than 21,000 people came to know Jesus Christ as their Savior because of this wonderful miracle of resurrection.[5]

After Timor, we left for the Philippines. When we told missionary Ernie Reb, who is doing new tribes evangelism, about the resurrections on Timor, he smiled. "I can believe this. We've had a number of resurrections in the Philippines."

Beyond and back experiences are being reported all over the world. Recently Evangelist Dick Mills, who is part of the ministry at Melodyland, met a man in South America who had been dead more than two hours. The man's brother, who was making the funeral arrangements,, and other witnesses confirmed the death.

The area of Micaway, Nicaragua, was extremely hard to penetrate with the Gospel. Christians were fasting and praying for the town, knowing it would take something drastic to reach the people.

While the brother was away making funeral arrangements, missionary Allan Tolle took authority over death in the name of Jesus. The man came back to life, and as a result more than 200 persons decided to follow Christ.

In Mexico one of the most startling resurrections happened in Jarretaderes, a village 10 miles north of Puerto Vallarta.

Phil and Doris Smalley, lay people from Melodyland, were used of God in this miracle. Phil, an usher at Melodyland, has car rental agencies in three Orange County cities. Doris is a secretary for an aerospace company.

Doris Smalley with two Mexican children in front of Margarita's palaypa in Jerretaderas, Mexico.

In January 1977, they flew to Mexico to visit with friends. Here's how Phil relates his part of the story:

> In Puerto Vallarta, we met Juan Castillo, a bartender I knew during my drinking days. He invited us to his house. He and his wife, Luisa, have five children. One of them, a three-month-old daughter, was blind in one eye. We prayed for her, hoping that God would provide a miracle to lead these people to Christ.
>
> But God had a bigger plan. Through a series of events we learned that Luisa's, grandmother in Jarretaderes was near death. By the time we arrived she had died. A relative had already gone to Puerto Vallarta to buy a casket.
>
> I had a small instamatic camera with a flashbar, and some of the family wanted a picture of the dead woman, Margarita Rodriguez. We went into Margarita's palaypa and found her lying on a bed at the far end of the hut. Her face was covered, her arms folded and a large crucifix lay on her chest.
>
> The Palaypa is a long room made of skinny sticks. It had a corrugated metal roof. With no windows or electricity, light filtered in between the cracks of the walls and through the small open door. The hut had a dirt floor, and chickens were free to roam about the room. About 30 mourners were jammed into the stick house, watching me take the picture.
>
> They removed the cover on Margarita's face, and as Doris chased the flies away, I took her picture. The smell of death filled the room so badly we could hardly stand it. She had been declared dead by Dr. Javier Arturo Mejorada Madrigal.[6] As we looked at Margarita, we noted that her eyes were set and her body was rigid and cold. Rigor mortis had set in, and the family said she had been dead for about six hours.
>
> Hoping to give some comfort to the mourners, Doris turned to the 23rd Psalm and began reading—in English, for

she cannot speak Spanish. None of the family could understand English.

Suddenly Doris had a deep sense of God's presence, and almost involuntarily placed a hand on the dead grandmother's head while the other hand holding the Bible flew into the air. She began to pray in a language she had not learned.[7] But the people recognized it as Spanish.

Doris looked down at Margarita then took a second look. Was the woman coming back to life?

Mrs. Smalley continues the story at this point:

I saw her nostrils moving and realized she was breathing. Within five minutes her chest started heaving. At this point the family crowded around the bed, staring at Margarita, their mouths wide open. Her stomach started moving in and out with her breathing. One of the daughters took off the cover and felt her legs, which had been stiff.

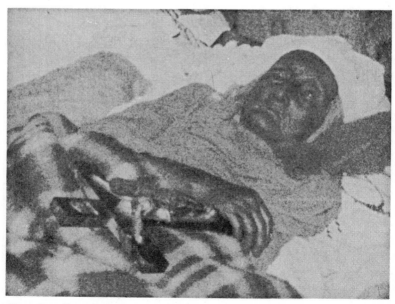

Grandmother Margarita Rodriguez who died and came back, living 24 hours.

I also touched her legs and could feel the life coming back into them. They were getting softer and returning to their natural color.

Still praying in the Spirit, I took Margarita in my arms and told her how much Jesus loves her—in Spanish. Though I don't understand Spanish, I knew inwardly what I was saying.

Margarita sat up. Her eyes looked right into mine; they were shining like diamonds. Still speaking in Spanish, I told her to ask Jesus into her heart. She shook her head and said, "Si, si, si." She understood everything.

Her mouth was completely white. Phil brought a cup of water and a Kleenex to wash her mouth. Luisa got a spoon and fed her grandmother a whole cup of water.

Meanwhile Juan Castillo had entered the palaypa. "Oh, Phillipe, your wife, she speaks Español; she has much power" he exclaimed. "No, she doesn't speak Spanish; and she has no power," Phil answered. "That's Jesus in her."

"I want Jesus, too," Juan cried.

I carefully laid the newly resurrected woman down. "I heard that, Phil. Get a Spanish tract. The last page tells how to receive Jesus."

Phil distributed the tracts to the mourners, and they repeated the sinner's prayer over and over. "Si, si, it's Jesús it's spiritu," they cried. Juan rejoiced, "I felt Jesús come into my heart."

God began to heal. Sores and bites on their bodies began to disappear instantly. *Juan was healed of an eye problem and a crippled foot.*

Phil took another picture of Margarita, and we left for the beach to meet a friend. A short time later we returned and looked in on the grandmother. She had fallen asleep and was snoring. I noticed she was sweating from the heat, so we bathed her. Margarita lived for 24 hours then died. God had accomplished His purpose. Margarita had been beyond, back and beyond again.

In those 24 hours, she and her whole household had experienced a spiritual resurrection. They had made a commitment to Christ. Word had spread throughout the town that the God of miracles is still alive.[8]

Why Resurrections?

The credentials of the Gospel are miracles. And there is no greater need anywhere for miracles than in primitive areas.

A high mortality rate of 50 per cent among children exists in many lands. In remote mission areas it is either do or die; no other option or alternative is available. People have no Blue Cross or sophisticated hospitals to rely on.

The greater the demonic power, the greater the need for God's miraculous power. The greatest miracles in the New Testament were at Ephesus and Corinth, the most wicked cities in Asia Minor.

Does this mean God's resurrection power is not needed in America and other great nations? Do our intellectual pursuits and materialistic successes eliminate the need for the supernatural?

Absolutely no!

Today's sophisticated society demands the supernatural.

Do you know what it's like to leave your body and live to tell it? Denise Beck does. She was in an airplane crash and survived it. Rev. R. A. Work did. He went to Heaven and described it. Betty Malz does; she died and confirms it.

6

"I Saw a Pure White Light"

"Hang on! Brace yourself!" John Beck yelled.

The 24-year old pilot and his wife, Denise, watched breathlessly as the rough ground rose faster and faster beneath their light plane. The veins in his neck and forehead stuck out like little beads as John pulled back on the controls with all his strength, his white knuckles clenched in fear. Denise reached out to brace herself just as there was a tremendous thud.

The plane struck on its nose wheel, dug a shallow trench in the freshly plowed field, then bounced hard into the air and crashed again. Moments later, Denise was peering spellbound at a dazzling pure white light.

Mrs. Beck's account is only one of the fascinating beyond and back cases coming out of Melodyland. She told her story recently at a Thursday morning miracle service:

> December 26, 1976 was a beautiful day. There were no clouds in the sky. A perfect day for flying. And inside my husband, John, the call to the skies was too strong to resist.
>
> It had only been three weeks since we bought the plane, and we had been flying it often. John's father, his brother and uncle also were pilots, and they had all taken turns trying it out. John and I had been taking Johnny, our baby, up with us every time; he even had taken a ride with his daddy by himself a week before the accident.
>
> But today John called his mother early and asked, "Would you like to babysit for a while?"
>
> "Why?" she asked. He said, "Oh, I just thought it would be nice if Denise and I went up alone for once."
>
> We had just observed our second wedding anniversary, but with his work schedule, we hadn't much time to celebrate, so we were glad to be alone. At Meadowlark Airport in Huntington Beach, California, after a routine preflight of the aircraft, John taxied our single-engine, two-seater Ercoupe to the end of the runway, gunned the motor and took off. It was about a quarter to eleven.
>
> He then headed toward Corona (California) and was parallel to nearby Lake Elsinore when I heard a loud crack, like the popping of a whip. John realized something was wrong. Suddenly the engine quit, and the prop began windmilling. John began emergency procedures.
>
> We were always keeping an eye out for places to land in case of an emergency. The field we picked to land on was beautiful and just outside Murietta. There were no bumps; nothing to harm anything if we could just set the plane down.
>
> I sat back in my seat as far as I could force myself to lie back, and I could see the concentration on John's face. He

checked over every control and every instrument on the panel; he could find nothing visibly wrong.

We were sensible enough to remain calm and concentrate on the emergency. We knew we had to put the plane down. As the Ercoupe made its final approach to the field, John pulled back hard on the controls to keep the nose up and slow the plane down. The aircraft could not be stalled, a safety factor that had appealed to him when he bought the plane.

I could see the little veins in his neck and forehead sticking out as he pulled back with all his strength. I had to sit back on the seat harder so he could pull back all the way on the controls.

Just before impact John yelled, "Hang on! Brace yourself!" I remember reaching out and bracing myself with my right arm, then feeling a tremendous thud. The plane hit on its nose wheel, bounced into the air and crashed, ripping

The plane that John and Denise Beck were flying—after the fatal crash.

off the landing gear and settling on its undercarriage. The impact knocked us into the instrument panel.

Just before we slipped into unconsciousness, John turned toward me. His face reflecting deep concern, he reached out with his hand and touched the side of my face.

"Denise, I love . . . you," he said lovingly.

"I love you, too, John," I answered, dazed. Then John leaned over onto my lap, and within seconds everything was black around me.

I learned later from the doctors who cared for me that John died of a massive hemorrhage in the brain. I suffered fractured vertebrae in my neck, severe concussion just above my forehead, a fracture at the base of my skull, a broken right arm and a couple of fractures in my left ankle.

The clock on the plane broke on impact at 11:38. Rescue teams didn't reach the crash site until 1:42. The man who first discovered us thought we were dead. He was about to go for help when I cried, "Help me." He just about jumped out of his skin.

A strange darkness surrounded me as I slipped into un- consciousness. The darkness was extreme. I remember hearing a roaring, buzzing sound. There's nothing on Earth to com- pare it with. Then I saw a pure white light.

It was so bright it would make the sun look like a shadow. The light was everywhere, and the darkness was gone. I wasn't blinded by the light; it didn't hurt my eyes or make me blink.

I felt a tremedous love and happiness, an indescribable peace and joy and purity coming from the light. Then I saw John and Jesus.

Jesus' face was one no artist could capture. His hair was a

brownish auburn color, and it was to his shoulder. He wore a blue transparent robe over a lighter blue gown. Jesus was facing me, and John was standing in front of Him with his back to me. Jesus was almost a head taller than John.

My husband had his head buried between his shoulders and was telling Jesus how sorry he was for every wrong thing he had committed. He was begging Jesus to forgive him. I've never heard more sorrow in repentance.

Jesus spoke to him. His voice was tremendous and full of compassion. It had authority and power.

I could see part of myself as I stood there in the presence of the Lord. I was not in my body, but in spirit.

Suddenly, with the quickness of a blink, I saw my husband standing on the left side of Jesus. Jesus was holding his right hand, and I could see an overwhelmingly beautiful smile of happiness and love on John's face.

John wasn't wearing the clothes he had been flying in. He was wearing his three-piece blue-grey denim suit, a tie and the shirt he had purchased with the suit. He was standing in his stocking feet. I wondered why he didn't have his shoes on. About two weeks after the funeral I learned why. It is not customary to bury someone with shoes on.

The casket carrying the body of John Beck. Mrs. Beck in her out-of-the-body experience saw his funeral clothing.

Mrs. Beck, of course, wasn't able to attend her husband's funeral, but learned later that he had worn that suit for the service. Beck was wearing a flannel shirt and plaid denim pants when the plane crashed. She continues:

The smile that was on John's face was more beautiful than I have ever seen, even on our wedding day or when our son was born.

I was so happy that I wanted to stay right where I was. I didn't want to come back. But then Jesus told me I had to return. I was so disappointed. I didn't want to give up the love and the peace and happiness that was there. I was told to come back because of my son, Johnny.

Jesus said, "It will be hard, but don't worry. John will be happy with me here for eternity. Give all your love to God. And tell others that there is *life into eternal life* after death." Then Jesus began to explain the Bible to me.

As Jesus talked, the Bible suddenly began to become clear. Things difficult to understand, I began to realize their full meaning. I had been looking for something spiritual, but didn't know how to find it. I had not made a personal commitment to Christ even though John knew what it meant to be born again.

And when Jesus told me to give all my love to God, I felt exhilarated in the joy and happiness that was there. I said, "Great, I'll do anything you ask, Lord. Just tell me what you want me to do."

He told me to come back and raise the baby as God would want the baby raised and do everything the Bible says. Mom's first realization that I was now interested in spiritual things came during my second week in the hospital. One afternoon, I surprised her by saying, "Would you please do me a favor, Mom? Would you take me to Bible class with you when I get home?" Though still dazed by the tragedy, she answered joyfully, "I sure will!"

I got one final glimpse of John before my spirit re-entered my body. It was like blinking an eye, and the scene changed. John laid his head on Jesus' right shoulder with his face toward His neck. Jesus was looking down over him like a father looks at a small child. John had his eyes closed and was smiling.

The love I felt there was as though John, like a small child, was saying, "Here I am. Whatever you want me to do is fine." Jesus looked down at John and put His arms completely around his shoulders. The Lord's robe covered him to the bottom of his feet.

Suddenly the darkness returned, and when I came back from the beyond, I regained consciousness for a moment . . . long enough to say to a man who was bending over me, "Help me!"

Mrs. Beck was taken to Riverside General Hospital. But within two weeks she had amazed doctors with her rapid recovery. She was released to recuperate at home, her neck in a brace and casts on her arms and leg. But the story doesn't end here.

I would have lost my mind in the hospital had it not been for what I had seen and for what Jesus told me. The doctors and nurses couldn't get over how joyful I was. I knew John was dead. But knowing his happiness helped me to get well.

At home, however, I began to feel lonely, and doubts about my out-of-body experience began to crowd my thoughts.

I desperately wanted God to somehow let me know that what I had seen was real because I was being told that what I had seen was caused by my head injuries. I needed confirmation.

I felt hurt that I didn't get to the funeral. It seemed so

unfair that we were flying and suddenly he was gone, and I didn't get a chance to see him, to know that what I'd seen in Heaven was real.

Early one morning that strange blackness was there again. Being awake, I remember looking at the clock. It was about 6:30.

Suddenly, I was sitting on John's lap like I had always done if I needed a shoulder to cry on. We were sitting on a grassy area under a beautiful tree. I've never seen a tree like it. The leaves were perfect. We were beside a little stream. The water was like crystal. And that light was there.

I remember feeling his hands on my arm. Snuggled up to him like always, John told me what I had seen the day of the crash was true; he told me not to worry, that everything will be fine.

I was so happy, I wanted to stay there with him. I remember feeling his hand on my head, like he would always put it there, stroking my hair.

He assured me that we would be together again in eternity. Then that darkness returned. Before re-entering my body, I saw myself sitting up on John's lap. And when I came to, I was sitting about in the same position I had been while in John's arms.

Before her experience, Mrs. Beck says, she had a tremendous fear of death. But that's changed. She concluded her testimony:

I'm not afraid of dying now, though I was then. It's nothing to be afraid of. And I have a clearer understanding of God's Word. It's overwhelming! Often when I read the Bible, I realize, "Hey! That's just what Jesus said!" I know today that God is merciful in death.

When I went back in for my final checkup and the doctors

looked at the X-rays, they couldn't believe how I had healed. They were amazed at the difference between the before and after pictures, and they planed to use them in classes for medical students and technicians to show how my bones and neck weren't expected to heal the way they did.

The National Transportation and Safety Board determined after an investigation that the accident was pilot error due to low altitude. When Mrs. Beck came home from the hospital, her memory began to return and she recalled the cracking noise she had heard just before the plane's engine had failed. This prompted her father-in-law to demand another investigation.

Later, authorities determined the true cause of the accident. It was a faulty carburetor.

"In all my years of experience I've never seen a worse one," the investigator told the Becks. "How he ever got the plane off the runway, I'll never know. The carburetor had rusty water in it. The float was sticking, which produced too rich a mixture, and there was a dead spot in the carburetor creating a malfunction with the throttle."

You Must Return

Until recent years we seldom heard of resurrections or out-of-the-body experiences. Now the press is being flooded with these accounts from all over the world. I need not go outside my own family to relate a beyond and back trip. My father-in-law traveled this return journey.

Many times I heard my wife's father tell it from the pulpit to an electrified audience. At home it was our favorite miracle to repeat whenever the family or close friends got together.

My mother-in-law recalls it like this:

In March 1938 my husband, the Rev. R. A. Work, was pastoring in Shawnee, Oklahoma. A new building was under construction, and the builder asked my husband to come over to look at a window that was being installed.

Crossing a catwalk over some rafters, he stepped on a loose board and plunged headfirst nearly two stories onto the concrete floor of the basement among some pipes.

The workmen did not wait for an ambulance but rushed him to the offices of a well-known doctor nearby. One of the workers came to the house with the sad news.

"A bad accident has happened to your husband," he announced. "He's seriously injured, and they've taken him to the doctor's office. I came to take you there."

I slipped into a bedroom, closed the door and dropped on my knees. "Lord, I need you now. You said to ask and believe, and we would receive. In our years together in the ministry you have always been an ever present help in time of need. And Lord, I need you now. . . ."

Before I could finish, the Lord spoke into my heart, "I have heard your call. Your husband will be healed; there will be a miracle. But do not look at the injury. Keep looking at Me."

When I arrived at the doctor's office, they were dressing his head, and I was told to wait outside the door. But I kept walking right on into the room. The doctor said, "The Reverend has a serious concussion, a broken neck in the spinal column and a broken arm. In a few minutes he will be unconscious. So if you have anything to say to him, you had better talk fast."

My husband looked up at me and said, "You heard what the doctor said?"

"Yes, but I have talked to the Lord, and I have a promise that you will be healed."

Just then, as the doctor had warned, he blacked out.

"Mrs. Work," the doctor said kindly, "your husband probably won't last more than three days, but if he makes it that long there might be some hope."

"Well, praise the Lord! I'll be in Heaven," my husband announced.

"He's still conscious . . . better be careful what you say," cautioned one of the examiners.

"No, he is completely out. It's the subconscious mind reacting," replied the other doctor. "There is almost nothing we can do for him. The shock of setting the bones could bring on instant death. About all we can do is put him in the hospital and try to keep down the pain."

"But doctor, I have the promise that he will he healed!" I guess they thought I was beside myself. Anyway, instead of going to the hospital, I went home and began thanking God for his healing.

Early the next morning at the hospital a nurse came into my husband's room holding a needle. "What are you doing awake?" she scolded gently.

"Nurse, I have had a bright light in this room, and an angel has been with me all through the night."

"Oh, yes . . . but will you please do a favor for me?" she returned.

"I'll do a favor for you if you will promise to do one for me," he quipped back.

"Well, all right," she bargained, probably thinking to herself, "He's a dying man; he can't ask too much. . . ."

"Will you take the Gideon Bible from the drawer and read the 53rd chapter of Isaiah to me?"

Smiling faintly, she began leafing through the pages until

she located the Bible verse. When she finished the fifth verse, "And with his stripes we are healed," he interrupted. "Stop right there! I want you to pray for me."

"Oh . . . uh . . . ahem," she stalled, clearing her throat. "I've never prayed before anyone in public. Just privately at home, you see. . . ."

"Well, prayer is just talking to God. Whatever comes from the heart is all one has to say," he confided.

Bowing her head, she prayed a childlike prayer, but at the end she exclaimed, "You are going to make it. I just know you will be okay." She turned quickly and ran from the room.

The third day found me back at the hospital at my husband's bedside. As the doctor was standing there, both of us were a bit startled when he spoke up, "Doctor, I want to go home today. I want to check out of the hospital."

"But I have explained to your wife that the third day is the most crucial. You could either go into shock and die or lapse into a coma tonight. What would you do at home without a doctor or medical help?"

"Well, doctor, if I do die at home, at least I will die in a red hot prayer meeting! There is nothing more you can do here for my recovery."

Turning to me, the doctor again warned, "Mrs. Work, you may be taking your husband home to die. It's such a critical time period." But my heart told me to listen to my husband's request, so we left for home.

As soon as we got settled down at the house, he asked me, "When is the Lord going to heal me?" I said, "There is a time—I don't know when, but I have a promise."

I could see he was in misery and fighting the excruciating pain in his head and neck.

"Can you turn my head just a little?" he begged.

"I'll try." I barely touched his neck, when suddenly he gasped and passed out. We were staying in the home of friends, Jim and Tina Lawson. Tina was in the room at the time.

"My God, he's dying," she screamed and ran out the door to get some of the church people to come and pray. Again, I fell to my knees, rebuked death and reminded God of the promise He had given me.

"How can I ever know your voice again if this isn't you," I pleaded. At this moment, Bob's body jerked, and he opened his eyes. "Honey, am I dying?"

"You are not! You are healed," I answered joyously.

"But I saw Heaven," he said dreamily.

Some of the men had arrived and began joining in prayer. "Lift me up. I'm getting out of bed," my husband announced. The men helped raise him up to the side of the bed, but he sank back onto the pillows in weakness.

"Help me up again, and put my feet on the floor this time," he insisted. Again the effort failed, and he slumped back into bed. "I *must* get to my feet . . . please help me just one more time!"

The men pleaded, "Oh, no, Brother Work. You're just too weak. We can't put you through this kind of agony again."

But I moved to his side and said, *"Lord, this is the time."* It seemed I had barely touched him under the arms when his feet hit the floor. It was like new life had come into him. He started walking. He took one step, then another. He walked out of the bedroom, down the hall, then all through the house. He went from room to room shaking his head up and down and moving it around.

We were all beside ourselves with joy and praise. My husband was healed! God had been faithful to keep His promise to me.

Finally, Bob announced, "I'm hungry. I want to eat!" I believe that bowl of soup was the most fun I've ever had cooking. A short time later, he went back to his room and sat on the edge of the bed.

"I was with the Lord," he said quietly. "I rode in a heavenly vehicle through space toward a bright light. The light got brighter and brighter as the craft got closer. I stopped at the gate to a magnificent city with a beauty indescribable.

"I was surrounded by angels and loved ones and could see inside the city. I saw streets of gold and other splendors I have no words to describe. I could see people who had passed on to be with the Lord.

"Just as I was about to be ushered through the pearly gates, two heavenly beings stopped me. 'You can't go through,' they said. 'You must return. Your work is not finished.'

"But why do I have to go back?" I pleaded with them. One of the angels said, 'There's a prayer that *must* be answered. You *have* to go back.'

It was at that moment my husband's body jerked and he opened his eyes. The experience was so tremendous that for three days I would find him walking throughout the house in tears, reliving it.

"Honey," I would ask, "what are you crying about? The Lord has been so good to you. Why the tears?"

"Oh, I was right there," he would say. "I was ready to go. Why didn't you let me go on and be with my Lord?"

"Because we need you back; your work isn't finished on Earth."

"Yes, that's right," he would answer. "But it was glorious."

A few days later I drove him to the doctor's office. "What are you doing here? You should be in the hospital!" he exclaimed.

"I want you to take this stuff off my head. It's too heavy," he answered. He looked just like a mummy with the neck and head wrappings. The bandage started just above his eyebrows and was wrapped around his head and ears, the back of his neck and over his head. When the doctor unwrapped the bandage, he was amazed how the scalp wounds had healed so quickly. Only a small bandage was required.

The physician, who was a Christian, agreed that God had performed a miracle. Not only was it miraculous that he had lived, but the injuries had healed so fast.

It was impossible to determine if my father-in-law had been clinically dead or in the final stages of life, since no physician was present and sophisticated machines and monitors were not in use then. However, he often said that he had felt his spirit lift from his body as in death.

In 1960, 22 years later, my father-in-law was in another church building program. This time he was in Northern California. He came home one day with a heart attack, went to the bedroom and sat on the side of the bed.

Here's how my mother-in-law tells it:

I entered the room and realized what was happening. It just seemed that the Lord spoke to me and said, 'This is it.' We started praying and sent for the ambulance, but before it came, he was gone.

God had extended his life for 22 years. Our children were

married, and he had many more years of fruitful ministry.

Corpse Sits Up and Speaks

I'd like to share another resurrection case with you. **Dr. Walter Martin**, one of our professors at Melodyland School of Theology, told me about an incident that happened in the 1930s.

A woman of about 72 died in the Birchwood Nursing Home in Huntington Station, Long Island. She had no heartbeat or respiration. Her blood pressure was zero, eyes were dilated, and her bowels were released.

Attendants cleaned her and the bed, and the doctor pronounced her dead. The physician was an agnostic and made fun of the Gospel because it was a Christian nursing home. "Why bother me?" he laughed. "Your God is big enough to take care of the dead."

Pastor Carl Steffens, who was present, felt led to pray for the woman. In the presence of his wife and a nurse, he laid hands on the dead lady and asked God to raise her up. Nothing happened. So they covered her up, turned out the lights and left the room.

Four and a half hours later, the woman's son arrived. He went to her room with Pastor and Mrs. Steffens and the nurse and rolled the sheet back to get one last look at his mother before the undertaker came.

As he rolled back the sheet, his mother opened her eyes and sat up. "Albert . . ." she said.

Albert fainted.

Pastor Steffens staggered against the wall, and the nurse screamed. His wife nearly collapsed.

The doctor came back, examined the woman and realized

she had come back from the dead. But he never believed. He is in his 70s and still alive. This reaffirms the Bible verse where Jesus said, "Some will not believe though one be raised from the dead."

Child Comes Back to Life

Harry Rainville, who attends Melodyland regularly with his wife, Irene, is another person who has been beyond and back. Now 71 and a semi-retired golf course architect, he found the Lord at age 64, partly through the ministry of Melodyland.

His experience of death occurred when he was nearly three years old, living on a farm in the black hills of South Dakota.

Here's how he relates the story:

The farm was near St. Onge, a French community. My father was French Canadian and did his own blacksmith work. He had purchased a 75-pound anvil and set it on a stump in his shop. The anvil was to be anchored to the stump with horseshoes.

I was a peppy little fellow, and while Dad was standing at the forge heating an iron, I took hold of the point of the anvil and tried to throw my legs over it.

This upset the anvil, and I landed on my back with the point of the anvil almost through the middle of my stomach, rupturing part of my intestines.

A self-taught veterinarian, my father pulled the anvil off. My intestines were exposed, so he took them out, stacked them on my stomach and washed them. He then put the intestines back into place and wrapped my stomach with a cloth.

Fortunately a train was due at the town depot right after the accident, so the family rushed me by train to a hospital in Belle Fourche about 20 miles away.

I survived surgery, but got whooping cough and dysentery. In those days doctors didn't know how to control it, and I looked like a skeleton. I soon died, but Mother couldn't accept it.

"No, no! I don't believe he's dead," she insisted over and over. The doctor and his staff tried for nearly three hours to convince her there was no hope and that the body should be taken to a mortuary.

In a final effort to convince her, the doctor put a cold glass over my mouth to detect any sign of breath. He listened again for a heart beat and checked for pulse. Nothing. No sign of life. But she insisted her son wasn't going to die. The doctor and nurses finally gave up in disgust and left her alone with the body.

A Christian and strict Southern Baptist, she dropped to her knees and prayed. She prayed by my bedside all night. And God heard her.

About 9 o'clock the following morning, 17½ hours after I was pronounced dead, I began to breathe. I was nursed back to health, but my memory was gone. I had forgotten how to talk and walk. I had to start life over as a baby. From there I grew and matured normally into manhood and in later life became a golf course architect.

I Saw the River of Life

As a young girl Betty Baxter of Albuquerque, New Mexico, was hopelessly crippled and as deformed as the woman Jesus healed in Luke 13:10-13. She was miraculously healed, but not until she had been beyond and back.

She had been born with every vertebra in her spine out of place, the bones twisted and matted together. She also had St. Vitus' Dance, and shook violently from head to toe.

Even though doctors strapped her body to the bed day

and night, she didn't stop her shaking. The straps did keep her from falling out of bed. When they were removed, she would be raw and blistered.

Here are excerpts from her account:

> If I had not had Jesus, I could not have endured it. I was raised in a Christian home and gave my heart to the Lord when I was nine years old. Through the years I became worse and was given up to die. But I increased in faith.
>
> I had faith because Mom had taught me all about Jesus. And I had memorized her favorite Bible verses: "If thou canst believe, all things are possible to him that believeth," and "Nothing is impossible with God."
>
> During my years of loneliness, I became intimately acquainted with the Lord. Mom would bathe me in the mornings, then she would leave me. Sometimes I would hear a soft step by my bedside, and I would wonder if Mom had come in the room while I was not listening.
>
> Then I would hear the gentle voice that I had learned to know was Jesus speaking to me. He told me many things during those visits, but the one thing He always said that thrilled me most was, "Betty, I love you!"
>
> Large knots had grown on my spine; the first one at the base of my neck, then one right below the other to the base of my spine. My arms were paralyzed from my shoulders to my wrists. I could only move my fingers. My head was twisted and turned down on my chest. When I drank water I had to drink from a tube because I couldn't raise my head. Yet Jesus whispered that He loved me, even in this condition.
>
> As the years went by, I gave up all hope of getting well through medical care. But I still had faith that God would heal me. One day shortly before sunset I was struck with unbearable pain and lapsed into unconsciousness. On the fifth day, I regained consciousness, though I was still in severe pain and could not speak.

I began to pray to Jesus to come and take me to Heaven. As I prayed a thick darkness settled over me. I felt coldness creeping through my body. In a moment's time, I was cold all over and completely surrounded by darkness.

Through the darkness I saw a long, dark, narrow valley. As I went inside this valley, I began to scream. I knew this must be the Valley of Death. I had prayed to die, and to get to Jesus I would have to walk it. I started forward.

I had barely gotten inside when the place lit up with the light of day. I felt something strong and firm take hold of my hand. I didn't need to look. I knew it was the nail-scarred hand of the Son of God who had saved my soul.

He took my hand and held it tightly, and I went on through the valley. I wasn't afraid any more. I was happy, for now I was going home. My mother had said in Heaven I would have a new body, one that would be straight instead of bent and twisted and crippled.

At last we heard music in the distance, the most beautiful music I had ever heard. We quickened our steps. We came to a wide river separating us from that beautiful land. I looked on the other side and saw green grass, flowers of every color, beautiful flowers that would never die. I saw the river of life winding its way through the city of God. Standing on its banks was a company of those who had been redeemed by the blood of the Lamb, and they were singing.

At that very moment I heard Jesus say very softly and with great kindness, "No, Betty, it's not your time to cross yet. Go back, for you are going to have healing in the fall."

Instantly I returned to my body and slowly regained consciousness. God was true to His promise.

On a Sunday afternoon a few months later, she was miraculously and instantly healed. The Lord appeared to her, talked with her, laid His nail-scarred hands on her twisted spine. And in a moment her bent body was straightened and healed.

The incident happened in Fairmont, Minnesota, and the Fairmont Daily Sentinel carried the story of her healing in bold headlines on its front page. Many thousands have since heard her testimony and accepted Christ.

Of all the examples this is one with which I've not had personal contact, but I've heard her testimony before large audiences and consider it to be one of the greatest of our day. My friend Oral Roberts often had her relate the story in his crusades.

Every Morning Like Easter

Another remarkable case that's gained much publicity is the experience of Betty Malz of Pasadena, Texas. She says she had to die to learn how to live.

A piece of white paper is taped to her refrigerator door. On it is written "24 hours can turn any situation completely around." What she has to say in these next few pages is the exciting story of her journey beyond and back and the events leading to her experience:

It started as a glorious vacation. We had two weeks to spend in Florida. My father, mother, little brother; my husband, John; my daughter, Brenda Lynn, and myself. We had just arrived at Gulf Vista Retreat, north of Clearwater, Florida, and had spent a wonderful day on the sand and in the sun.

But about 11 o'clock that night, I experienced a pain in my right side. A pain I had never known before. My father and my husband rushed me to a little beach hospital in Tarpon Springs, Florida. Dr. James B. Thompson examined me and took a blood test.

"You have an appendix full and ready to burst!" he warned. "But we are in trouble. I have a 24-bed hospital here. I have 26 patients and two of these are mothers in labor and ready to have their babies. I'm the only doctor here, and I cannot operate on you."

So he called ahead to the Clearwater Hospital 22 miles away. They put me in an ambulance and sent me there.

The surgeon met me there and prepared me for surgery. Then he began to examine me. He punched and pulled and finally said, "I'm sorry, but I believe Dr. Thompson has made a bad analysis. There's not even any swelling in your side."

Then I remembered what Dr. Thompson told me: "If you get better on the way to the hospital, you'll know that the appendix has ruptured."

I stayed in that hospital for seven days. First they gave me Demerol, then morphine for pain because the pain did not diminish. I began running a high fever and to swell badly.

The doctors began to speculate on what really was wrong. One thought it was a ruptured appendix, the other thought it was just an infection from the heat and the strain of driving.

After this seven-day period, they decided to send me back home and let our family doctor examine me; perhaps we would feel more at ease with his diagnosis. After I went back home by jet, I lay there in the hot July weather for four more days. On the morning of the 11th day, I awakened, totally blind. I could not taste or smell. My fever was running from 103 to 105, and I was more swollen. The pain was unbearable.

They rushed me to the emergency room at the Union Hospital at Terre Haute, Indiana. Remembering that one doctor, upon examination earlier, suggested that I might have a four-month dead fetus from tubular pregnancy, they called in a gynecologist and a surgeon to operate on me.

I'll never forget the words of this doctor when I revived just a little after surgery.

"You had a ruptured appendix 11 days ago; you have survived the hot July weather. I found the nastiest mess I have ever cut into in my 28 years of practice. I found a mass of gangrene the size of a man's head, and it has coated all of the pelvic area and all of your organs. I have lost a 200-pound man in 48 hours. I don't understand how a 120-pound woman has lived 11 days."

They told me that I could not live, but I did live. I was there for 44 days, most of that time in a coma. Just twice, I regained consciousness, once when the pain became worse and they discovered I had a telescoped bowel, and another time when an abscess formed under the incision.

I had B-negative blood. Transfusions were hard to come by. And after many, many transfusions, because the gangrene would eat up the good blood, they began to struggle for my life. It was impossible to get more plasma in the immediate area. They called St. Louis and Indianapolis, got what they could, but they were running low on blood donors.

While I lay in a coma one day, I heard footsteps coming into my room. Soon I heard pages of a Bible being turned, and I heard the voice of a man that I never liked. He was an ugly man, crude and uneducated, and I was ashamed of him, but there he stood at the foot of my bed.

My first impulse was, "I can't stand this man when I am well. I know I can't take him. Doesn't he know I'm sick?" But he stood at my bed and began to read from the Word of God. He read perhaps two chapters in the Psalms. I'll never forget the one Bible verse he read: Psalms 107:20, "He sent His word and healed them."

All of a sudden I began to wonder, "Can I be healed?" A fountain of faith began to well up inside of me. I felt such a baptism of love come over me; it was like a warm shower, and I loved that man for bringing me the Word of God. I will never again judge a man for his appearance or lack of education.

God chooses the instrument that He will use, and that homely instrument played sweet music to my dying soul that day.

Soon pneumonia set in; I could no longer take intravenous feedings nor blood plasma because my veins collapsed. At this point, they called my family in and told them that they must all go home and get some rest.

At five o'clock the following morning my father was awakened suddenly. He felt compelled to go to the hospital, just to slip into my room quietly and to sit there with me.

Shortly after he had left, the hospital telephoned. They first called my husband, then they called my mother and told her that I was dead. Could the family come early that morning at their convenience and make arrangements for my body.

My mother said after the first initial shock she remembered my father was already on the way to my room, and she asked them to please intercept him.

My father says he will never forget the sight when he slipped in the exit door and quietly went into my room that morning. He realized that something had happened because all the equipment was removed and a sheet was pulled over my face.

He says that the despair, the disappointment of that minute in my room frightened him so, he did not know what to say or what to do. Though he was a minister, he was at a loss for words. It was even too late to pray for me, but he says he felt he must pray for strength for himself. He bowed his head and all he could think of was, "Jesus." He said it twice, "Jesus, Jesus."

As he said, "Jesus," I heard his voice; I also was hearing the echo of the most harmonious, melodious voices. I remember walking up a beautiful green hill; my legs were not tired and the moving was effortless.

Beside me was the most majestic silver marble wall, and I found myself at the top of this hill. The grass was green and soft as velvet, and each blade seemed alive. The voices were singing, and I began to sing too.

I have always had a girl's body but a boy's voice. I've always wanted to sing high and sweetly. All of a sudden I began to sing high worshipful tones, and they blended with those coming over the wall.

Beside me I saw the skirt of a white garment. This person, who must have been an angel, stepped forward, pressing the palm of his hand against the most beautiful translucent pearl gates with gothic scrolls at the top.

There were no handles on these gates, and as he pressed against them he asked me, "Would you like to go in and join in the singing?" I answered, "No, I would like to sing from here, and then I would like to go back to my family and back down the hill."

He nodded and stepped back. I stood there and received something I can never explain, but the majesty of the presence of Almighty God in Heaven is something I shall never forget. I turned around and started down the same beautiful green hill. Over this silver marble wall I saw the most magnificent sunrise I have ever witnessed.

It actually was morning. It really was sunrise, and the rays of the sun were coming into my hospital room. I found myself feeling the warmth of those sunbeams. It penetrated the sheet. I felt the warmth coming into my body; I saw a sunbeam coming through the top pane of the glass from the window to the right of my bed. The lower part was covered with an air conditioner, but the top panel of glass was a clear pane, and as that sunbeam began to slant across my bed, I saw through the center of it ivory letters of about two inches high.

They poured down the center of the sunbeam, similar to a stock market tape report. They were John 11:25, "I am the resurrection and the life and he that believeth in me though he were dead yet shall he live."

I reached up to touch the Word of God that I saw coming through the sunbeam. When I reached up, I pushed the sheet off my face, grasped those words, and when I touched them, warmth and strength and life went into my fingers down through my arms and into my body, and I sat up!

No man can claim the credit for this experience. It was the Word of God that became alive to me. It was nothing I earned. It was life; His Word healed me. He sent His Word and healed me. Now, every morning is like Easter.

Running Scared

Several years ago I spoke at the dedication of a new church facility in Togo, West Africa. The missionary pastor, Bill Lovick, also operated a Bible School for training native evangelists and workers. I vividly recall him telling me about one of his students.

Coming from a remote tribal area, this pastor had enrolled in the school. One of the requirements was that students learn French proficiently enough for preaching, since it is widely used throughout Africa.

This young native could not grasp the language, so he met with Missionary Lovick and in a beautiful spirit left the school. Though he could not learn to preach in French, he reasoned, he could still minister in his dialect to his very own tribe.

Maybe, after all, that was God's plan for him.

Back in his tiny village, he was standing in front of his church one morning when he noticed a funeral procession coming down the road. By the time it reached the church he had learned the procession was that of a seven-year-old lad, the only child of a widow.

In Africa a widow traditionally lives with a son. To be left not only a widow without a son, but also childless was a compounded tragedy.

The native pastor later related how he felt a sudden compulsion to halt the procession. "But God, if nothing happens I will be an outcast here forever," he breathed.

Nevertheless with New Testament boldness, he stepped forth and said to the bearers of the corpse, "Bring the little boy's body inside the church and place it on the altar."

The natives obeyed immediately. The church was packed as the villagers streamed inside the building, wonder and curiosity on their faces.

The preacher stepped up to the altar. He described a rush of power and authority. At that moment, he related, he sensed the overwhelming power of the resurrection of Jesus, plus all the authority that had been given the 70. He uttered a simple command, "In the name of Jesus, rise."

The little boy stood up . . . instantly. The natives jumped up, ran out of the building and headed for the bush. Filled with fear and superstition, they were running scared.

The mother joyfully returned home with her son. The sensational news spread through the village, and revival broke out with almost the entire population turning to Christ. *One* miracle was the key. The God of miracles had done what the witchdoctor and voodoos could not.

While this is a valid case of a pastor commanding the dead to rise, what of those who attempt to bring back the dead and fail? Is praying to bring back the dead faith or fanaticism? This will be examined in the next chapter.

7

"Wesley Will Return to Us" —But He Didn't

"In the name of Jesus Christ," thundered Lawrence Parker. "Death, loose this body. In the name of Jesus Christ, arise!"

The chubby man of 34 looked down expectantly at the body of his 11-year-old son Wesley laid out in a satin-lined, pine coffin in the steamy little funeral parlor in Barstow, California. The lid was open, Wesley's eyes were shut and his hands were clasped over his favorite brown sweater.

Wesley's two sisters and brother stood eagerly on tiptoe beside his mother watching his father peer into the coffin, his Bible in hand, opened to the Gospel of John.

"In the name of Jesus, arise!"

But nothing happened.

Parker, an unemployed aerospace worker, and his wife,

Alice, had come to Barstow in 1962 with their infant son, Wesley, the first of four children. A quiet and God-fearing man, he attended a small church with his family. His faith carried him down byways that even his pastor thought bizarre.

Wesley was a friendly boy, popular with his schoolmates, but he tired easily. At age six, doctors discovered that he was diabetic. Not quite satisfied with the diagnosis, the Parkers nevertheless tolerated a course of insulin treatments for the boy. Occasionally the boy's condition would improve. Thinking it was an answer to their prayers for healing, they would take Wesley off his insulin—only to have the symptoms return.

Then in the late summer of 1973 an evangelist held a service at the Parkers' church, and they took Wesley forward after the sermon for a miracle. He anointed the boy with oil, placed hands on his head and prayed. Those in the meeting believed Wesley was healed.

Later, the Parkers scrawled the good news in Wesley's

Lawrence Parker, clutching Bible, is led by police to a patrol car. Charged with involuntary manslaughter, they were accused of withholding their son's insulin on grounds of religious belief. (United Press International photo)

insulin log: "Praise God, our son is healed," while the boy ran happily around the neighborhood telling friends.

The next morning, however, when Wesley tested his blood sugar, the results were still positive. He was just getting out his insulin kit and hypodermic syringe when the father grabbed the works, squirted out the insulin and broke off the hypodermic needle. "This is a lie of Satan, Wesley; we're not going to believe that!" he thundered.

Not long afterward, Wesley developed the usual symptoms—nausea, then agonizing stomach cramps. Their pastor and several friends were called to pray for the sick child. Parker later claimed he had personally exorcised a demon. Wesley, however, fell into a coma and died within three days.

Taking the defeat only as a momentary setback, he reassured police when they came to investigate, "Wesley's going to be resurrected." With that, Parker scheduled his resurrection service, although allowing the mortician to have the body embalmed.

"Christ is going to have to replace that blood anyway," Mrs. Parker said. "It was full of sugar. So it might as well be embalming fluid."

At first, the Parkers predicted Wesley would come back to life during the resurrection service. But when he did not, they went back to the Bible and calculated that their son, like Lazarus, would rise only after four days in the grave.

The father had the mortician bury his son in an unmarked grave, with only the gravedigger and the undertaker present. But the fourth day came and went, and conceding only to a "little disappointment here," the parents settled in for a long wait.

For a long while the Parkers kept their hope alive—even as the local district attorney ordered them to jail on a manslaughter charge and held them on $10,000 bail each.

"Wesley will be returned to us," they said. "God *must* honor His Word. It's as simple as that."[1]

Alice and Lawrence Parker leaving court after the pretrial hearing on the death of their 11-year-old son, Wesley. (United Press International photo)

Dick Mills, an evangelist and teacher at Melodyland, met the Parkers shortly after the incident. He relates:

> I was in services in Victorville, about 35 miles from Barstow. Mr. and Mrs. Parker were in the audience, but I did not know they were there. At that time there was extreme pressure upon them as they waited trial. Often they would come and visit the church in Victorville.

Mills has an unusual gift of the word of knowledge, which

he uses in connection with verses from the Bible. The word of knowledge is the spirit of knowing. It's a supernatural gift that reveals certain things to the Christian believer (I Corinthians 12, 14). A counterfeit to this gift is ESP.

Usually Mills does not know how the Bible verses he uses relate to the person he ministers to. When he gives a word, it usually is one of comfort and strength; it is not vindictive or judgmental. Mills had such a ministry to the Parkers that night.

I felt led to call this couple out of the audience. I asked them to stand up, then told them they were holding on to an idea that was impractical. "You're clinging to something," I said. "You are going at it the wrong way."

After the service was over, the Parkers came to me and asked, "Do you know who we are?" "No," I said. "We are the Parkers. We are the ones whose son died. Do you feel that we failed God?"

"It's not my place to judge, but you imposed your will on the boy. You really didn't take the boy into prayerful consultation. You announced to him.

"I have two verses for you. One is in Isaiah 57, 'The righteous are spared from the evil to come.' If the Lord takes them away, He doesn't intend to bring them back.

"Then here is another Bible verse (at this point I did not believe that the Parkers accepted the verse). The setting of the Scripture is where David's illegitimate child died. 'Why should I fast when he is dead. I cannot bring him back, but I can go to where he is' (II Samuel 12:23). That boy cannot come back, but you can go to where he is, Mr. and Mrs. Parker."

Mr. Parker resisted. "That boy *is* going to be raised from the dead!"

"When blood goes out of the body, the life factor is gone,

for the Bible says that the life of the flesh is in the blood,"
I countered. "However, if God wanted to turn embalming
fluid into blood He could do it. He turned water into wine."

Later, I dealt with the Parkers about motives, quoting Jere-
miah 45:5, "Are you seeking great things for yourself? Do
not seek them. . . ." Again the Parkers resisted this.

The Parkers admitted their error in a magazine article
just before this book went to press. "The faith that we were
attempting to exercise was the wrong kind of faith," he wrote.
"We wanted to see our son healed, but went about it the
wrong way. In withholding medicine from Wesley after he
was prayed for . . . we condemned our son to an early grave.
Through our experiences we have learned that medicine is
good."[2]

Since the printing of this article, I flew to Barstow to talk
with the Parkers. He admitted that for three years they believed Wesley would still be resurrected. I sensed a sincere
sorrow for their previous extreme position. Parker added,
"We had a lack of knowledge but no wrong motives in what
we did." I sensed no animosity in them, only a concern that
other people would not repeat their mistake. Parker insisted
their sole motive was, "We wanted our boy back."

In another incident, a frenzied sect in Arizona hauled the
body of Otis Harrell, a 75-year-old grandpa, from one Indian
reservation to another while drawing on his railroad pension.

Harrell, whose decomposed remains eventually were discovered hidden behind a false wall of a tool shed in Payson,
had died in October, 1975. His body had been kept at the
Phoenix Light Temple for about two weeks before it was
moved to another location in the city. Later, Dorothy Bell,
pastor of the small church and Harrell's daughter, began moving the rotting body around the state while they and members
of the congregation prayed in an attempt to bring him back

to life. After a while, the group bought a home in Payson, and the body was moved there.[3]

Before this incident, Mrs. Wilkerson and I went to Phoenix to consider purchasing property for another Melodyland work. The land adjoining this piece belonged to Phoenix Light Temple. At that time the Bells were also interested in selling their facility. Mrs. Bell introduced me to her father, Grandpa Harrell. He was sitting in the house, which adjoined a little chapel. Later, we decided against the purchase.

It was through this business contact that we sensed the extreme teachings of these people. It didn't surprise me when I read the newspaper account of their attempts to raise Grandpa Harrell from the dead.

In December 1976, the *Los Angeles Times* carried a similar account. For two months the unembalmed corpse of a 29-year-old man lay in a bed in a New York apartment while six of his friends prayed over the body, hoping to bring it back to life. Neighbors reported the odor of strong incense coming from the room, and shortly afterward police found the friends grouped around his body, chanting, "Rise, Stephan, rise!"

Later the six men said they were brought together through their "relationship with a man who had healing powers and knowledge about health matters of an unorthodox nature."[4]

This situation had nothing to do with the church; these six individuals simply believed they had the power to bring back the dead. Sometimes people blame Christianity for producing extreme cases, but extremists come from every walk of life.

Sovereign Acts of God

God is not an errand boy, although some people seem to have this idea. There is a verse in the Bible that says, "Command ye the hands of the Lord." But if you look at the context, in no way does it indicate we can control God. A lot of people today have an idea they can force God to perform miracles. Bringing back someone from the dead is exclusively a sovereign act of God. You cannot say, "Look, God, you have to do it because I have to prove your power."

There's not one promise in the Bible that tells us to resurrect the dead. In sending out His 70 disciples, Jesus commissioned them, "Raise the dead." However, that command has not been given directly to the church.

On occasions in early church history and in modern times, God has raised people from the dead. And while today there is Scriptural authority to believe God can and will raise the dead, there is no direct promise saying *He must do so.*

Resurrections Must Glorify God

A missionary couple from Mexico was stunned when two of their children drowned in a swift river. They refused burial for five days believing the children would be raised from the dead.

The children's bodies were taken to a funeral home in Garden Grove, California. I was telephoned to counsel with the parents, and they came to my office.

These dear strangers quoted the verse, "If *two* or *three* agree . . ." and said, "Pastor, we are going on God's formula. The third person will make their resurrection more certain." Immediately I discerned this was not God's plan. So I said no.

A few years later I learned that these missionaries admitted they thought they had super power.

Remember Mother Sarlin in Indonesia, whom God used to raise at least two people from the dead? Because she began taking money for performing miracles, God did not get the glory. So He withdrew His power from her.

Resurrection Should Spread the Gospel

A miracle is never given for man's gratification. It has a

purpose. As we have already seen in a previous chapter, the miraculous was given to authenticate the Gospel and produce great climates of faith.

When I talked to the Christians in Indonesia, they were amazed that I was interested in those who had come back from the dead. They said to me, "Why, Pastor, what's really glorious is that whole cities are turning to God." To this I readily agreed. The great miracle is not that a body is raised, but that those who are dead in sin have been resurrected to eternal life in Christ.

The Parker case brought disrepute to the supernatural. Again I say, if a doctor puts you on medication, he alone should be the one who takes you off it.

One of the most famous evangelists who prayed for the sick was sued for practicing medicine without a license during a crusade. He told a seven-year-old boy, who was a polio victim, to remove his brace. Under the law of that state, it was interpreted that the evangelist, who was in his thirties, was prescribing instead of praying for the sick.

Again I emphasize the importance of a physician attesting one's healing, for it strengthens the testimony.

Ironically, before the case got to court, the evangelist himself died of polio. Shortly before he took sick, I met him in the Bahamas while conducting a missionary crusade.

The evangelist was meeting with his lawyer on the case. At that time he admitted he had made a mistake in ordering the child to take off the brace. This was a head mistake, not one of the heart.

Resurrections are Temporary

Even if you die and are brought back to tell about it, death is only postponed. It's still ahead until the day of the great resurrection. All who have come back to life today will die again unless they are caught up to be with Christ at His return when He comes for His church.

Faith vs. Fanaticism

Sometimes there is a fine line between faith and fanaticism.

My heart goes out to the Parkers in their misguided faith, and I write with all the tenderness I have. Only a person facing the hour of tragedy can feel the pain and disappointment because there was no miracle. I give them my sincere sympathy.

Nowhere does the Bible teach that Christian healing is adverse to medical science. Many physicians who attend Melodyland services believe in divine healing. I have great esteem for the medical profession. No one that I know who practices Christian healing is against the medical doctor. If one is healed, his physician will know it. Because it's factual and true, what God does stands up under any scrutiny.

I remember so well the Baptist doctor I worked for when I was a student. Sometimes he would say to me, "I don't know the answer to this problem, Ralph. But you pray with me, and we're going to ask God to perform a miracle." Because of this atmosphere, I saw that the medical profession and the ministry could work side by side.

I've had people say, "I wouldn't go to a doctor because I don't know what's wrong with me, and I don't *want* to know what's wrong." But that's ridiculous.

Personally, I feel one needs an examination at least twice a year; you could find out that you're in better health than you thought!

Many people have been confused in an hour of sickness when some unwise person has said, "Brother, you are healed." If Wesley had been cured, this would have been confirmed had they taken him to the doctor. But the boy wasn't healed, and the Parkers' disregard for the medical profession cost the boy's life.

Another error is attributing everything to Satan. Usually this instruction does not come from the church, but from some little fringe group or false teaching on a tape.

There's been so much overemphasis on the teaching of deliverance from devils that many people live in constant fear. I've heard of cases where people will not get out of bed until someone calls and exorcises them.

So much fanatical teaching has spread around the country that I like the title of one book, *I Have Been Delivered from Deliverance.*

Christians can be oppressed, buffeted and attacked by Satan, but "greater is He that is within you than he that is in the world."

Extreme teachings can bring harm to the faith life of a church. In the next chapter, I will deal with several aspects of Christian healing and answer such controversial questions as *Why isn't everyone healed?* and *Why do men and women of miracles die?*

8
Every Healing a Minor Resurrection

"Tiny Lacinda Watson played happily at her Merewether home today unaware that she had made medical history as the youngest Australian to have an animal heart valve transplant," the Australian afternoon newspaper reported.[1]

"Surgeons at Prince Henry Hospital, Sydney, save two-year-old Lacinda's life by providing her with a heart valve cut from a pig.

"Today, little more than a month after the operation, Lacinda is a new person. She has gained weight rapidly and is able to live the life of a normal, healthy child."

This amazing story is the fulfillment of a prophetic word given by the Holy Spirit to the parents of this little girl the day they brought her to a Melodyland miracle service.

I learned about Lacinda's recovery in a letter I received from her grandmother.

Our granddaughter was born on January 11, 1974 and it was discovered soon after that she was suffering from congenital mitral incompetence and would need a mitral replacement to the heart to survive.

She had had heart failure since the age of one week. A cardiac catheterization at the age of three months confirmed the diagnosis.

About the first of November 1975 the parents, my son and his wife, were told that an operation was scheduled for the next week because the doctors could not wait any longer. Even though they feared she would not recover, they had to do something.

Judy (Lacinda's mother) requested that they at least have Christmas together, then do the operation. The child's chances of making it were very slim; she had weighed 17 pounds at 5 months, and at 23 months was still the same weight.

It was at this stage that my husband and I suggested that they should read Kathryn Kuhlman's book, *God Can Do It Again,* which we had given them eight months before. They did, and within the next week, after much prayer and consideration, we headed for Los Angeles.

We attended the service at the Shrine Auditorium on Sunday the 16th of November and expected a miracle. On Thursday our friends took us to Melodyland, where you were conducting a prayer and healing service.

It was there the power of God came upon Lacinda. Her parents gave their lives to Christ. The very next day the physical shape of her chest changed, and we knew God had touched and healed her.

You predicted that her healing would make national headline news.

When we arrived back in Australia, the doctors agreed her chest had changed shape, but the operation was still scheduled for February 4, 1976. It was on this day that God gave her a new heart, using surgeons to complete the miracle.

Lacinda is just one of thousands who is being helped by the power of Christ in the miracle services. And her case illustrates how miraculous healing and the medical profession can work hand in hand.

All healing is divine, but God does not always work in the same way. Some recoveries are instant, others progressive; many have been healed without further medical help and have gone to their doctors only for confirmation. But often God uses a physician or surgeon to complete His work. This is why I work closely with doctors.

Often I have been asked, "Why do many who come to miracle services go away without being healed?" "Why do men and women of miracles, like Kathryn Kuhlman, die?" And, "Why are only a few people resurrected?"

In this chapter I will deal with these and other vital issues of Christian healing.

But first, I want to share with you some other exciting testimonies. One is from a musician, James Isaacs, of Cerritos, California:

I am writing to share with you a healing that took place at Melodyland recently. I was the recipient.

A couple of weeks prior to this, I had been infected with a swollen runny sore at the back of my neck. A doctor at Kaiser Hospital in Bellflower advised me to see a specialist and have some tests run, which I did.

The next week, I had a regularly scheduled annual checkup with a government doctor, who also advised me to see a specialist.

In the meantime, this sore was getting worse, and I had to spend a lot of extra time each day dressing and bathing it in solutions.

Soon the test results came back, and it was diagnosed as a serious form of cancer for which there is no known cure. However, I did not cling to it because I have learned an awful lot about healing these past few years at Melodyland.

When the visiting preacher spoke that night, that was the climax to my faith experience. He gave a vivid message concerning the time when he was on his deathbed and had an outside the body experience. He related also that when he returned to the physical realm, you were there laying hands on him and praying for him.

Well, after the service that night, I went back to the special room for prayer; my wife went with me. I named the disease (as diagnosed) to the preacher, and he laid hands on me, cursing the ailment by name, in the name of Jesus.

About three or four days later there was no evidence that there was ever a festering sore, and when I returned to the doctors, they were amazed.[2]

During World War II, Don Allen suffered a severe head injury. He was thrown about fifteen feet by a bomb blast, which fractured his skull from ear to ear. Spending a minimum of time in a government hospital recuperating, Don didn't think much about his injury until thirteen years later when his wife came home from church on Sunday morning and found him draped over a chair, critically ill.

The first time I met Don was when I visited him in the Veteran's Hospital, although his sons were members of Melodyland. I prayed for him, but his healing did not occur until sometime later at a service.

Here are excerpts from his story:

It was close to Christmas time when my wife came home from church and found me draped over the chair. She took me to the hospital, and the doctors said, "It's his heart."

So they admitted me, but after a while, I was up and going again. They told me to take it easy and gave me certain medicines to take.

At this time I was a successful contractor in Los Angeles. We were doing probably 150 to 200 homes under contract every year. But strange things began to happen.

I would come out of a building, and it was like coming into a new world. I wouldn't know how I got there, what I had driven, where I had parked the car, the truck, the horse, or whatever I came in.

I would sign contracts not knowing what was in them. My brother finally had to take over the business.

It got worse until one day while my wife and I were getting ready to go to a funeral, she heard a thud in the master bedroom. There I lay, between the bathroom and the bedroom, paralyzed from the waist down.

My wife rushed me to the hospital, where the doctors said, "He's had a stroke; he hit his back on the door jam going down, and he paralyzed himself. It'll be okay." Sure enough it was. In a short time I was back on my feet again, but these strange things kept continuing.

Finally, the doctor said, "We've got to get you into someplace where they can constantly watch you." So I went into Veteran's Hospital, and there I started to deteriorate.

I was in and out of that hospital for the next five years. A calcium growth had started inside that old fracture, which cut off the blood supply to the left side of the brain. My brain started to waste away until five-eighths of it was gone.

Over the years I developed psychomotor epilepsy, and I was taking heavy medication . . . enough to keep a horse asleep, the doctor said. But it would just barely slow me down. My brain was running at such a rate it was like a short-circuited motor. It just burned itself up.

I developed an ulcer. The top of my stomach was gone, eaten away. Because I could bleed to death in just a few moments, someone had to be with me all the time.

One hand was like a claw and was more or less dead, while the other wasn't much better. My right foot had a drop; speech was slow and slurred. I had no feeling; and to get something, I would have to reach and know I had it.

I would have a seizure and fall. If I scratched myself, the wound wouldn't become infected; the skin would just dry up and flake off.

I had turned my back on the Lord about 15 years before, and when people would come and pray with me, I'd say, "When I left Jesus, I was a man. I was master of my ship, and when I get back that way, I'll rededicate my life."

One night I went to Melodyland Christian Center with a friend of mine. During the invitation, I was holding onto the pew and he was standing next to me. My friend started to move toward the aisle, and I stepped back.

Something inside of me said, "Don, there's one thing you've never done, and that is to stand before man in failure." So I started down that aisle to rededicate my life to the Lord.

The next Sunday night I was back in church. We were standing ready to be dismissed, when Pastor Wilkerson said, "I feel that the Holy Spirit would have me to change the order of the service; that He wants to do something for someone here tonight.

I don't know how, but I knew it was me. Pastor Wilkerson had visited me in the hospital, so he knew my condition.

When I made it to the front, the elders of the church prayed for me.

The following Wednesday I was in my garage working on some furniture when all of a sudden I felt something in my dead hand that resembled a claw. I turned that old hand over, and there was a sliver of wood.

It was the first time in five years that I felt pain anywhere in my arms or legs, and I ran through the house shouting, "The Lord God has healed me!" Within 30 days I was back to work—climbing walls and over roofs, cutting wood and picking up heavy saws—just like I had never quit!

Jesus had done more for me in fifteen seconds than the total medical resources of the universe could do in seven years!

I went back to the hospital sometime in March for a check-up by the doctor whom I had seen weep many times because there was nothing he could do for me.

The doctor said, "It doesn't appear that you need surgery at this time, Mr. Allen." I said, "I know that. Jesus healed me!"

He patted me on the head and figured, "That's natural coming from a guy with half his brain gone." That would be a natural thing to say. "You go home and take it easy, and everything is going to be all right."

I had been given only two years to live. Each time when I went in for an examination, all they wanted to do was up my pension and send me home. They couldn't believe it.

Finally I went to the Veteran's Administration building in Los Angeles for an examination. They would look at the X-rays, then at me; they would examine me, then read the charts.

"Is this you?" "That's me; that's me!" I answered. "You've taken medicine?" "Not if I could keep from it," I returned. On and on they asked questions, but I didn't tell them Jesus had healed me. At four o'clock that afternoon I was seeing the psychiatrist, trying to explain what had happened.

Finally, they were convinced I was cured, so they took away my pension and gave me a letter saying I have no disability.

God has performed many healings during the annual charismatic clinics at Melodyland. Mary Roland, a lady attending the services from Phoenix, Arizona, wrote me about the miracles she and her son experienced:

On Saturday night, both my son and I received a marvelous healing of the Lord. He was totally deaf in one ear and about half deaf in the other. Also, he was losing his sight. God healed him of both.

The Lord also gave me a new hip socket that had been destroyed by arthritis. When the healing took place, a burning sensation went through my body.

Dick Humphreys, a member of Melodyland, is another who received a miracle during a clinic. On his eleventh wedding anniversary he had a severe accident, which caused him and his family suffering for 15 years. Here's how he tells it:

I was setting chockers in a logging operation when a cat-skinner pushed a redwood sucker, about two and a half feet thick, over to make room for his skid row. It became hung up on a root and fell sideways.

Where it hit me, it was about nine inches thick. Because I was atop a big log, the hit was a banging blow instead of a crushing one. The tree bounced, and I was hurled a long way into the brush.

I suffered 14 fractures in nine ribs, which mended eventually; liver contusions, which healed naturally; multiple lung lining tears from which I had to be rescued twice by the doctor inserting needles to draw off blood and pus; the fracture of some vertebrae and the residual dysgenic disease that left me with pain from my neck to my ankles.

Surgery was out as far as I was concerned. The prognosis was pain, paralysis or cure.

I spent the next two and a half years in and out of the hospital, while our family of six suffered in numerous ways with me.

Until recently, I had not spent a day without pain since the accident. During those 15 years, my injuries caused not only pain, but inconvenience and loss of income.

But now to the exciting, yet mysterious part of my story.

What happened that night in August[3] is simply supernatural, and glory can only come to God because of it. I had no faith. I was not looking for healing. But when people with back problems were asked to stand, I stood.

At that moment I felt a physical tingling wash over my body. I didn't recognize that the Holy Spirit was at work physically because I did not believe it was possible. I experienced more freedom of movement, bending, twisting, at that time. But I continued to experience severe pain.

Early in November I suffered extreme pain, with no apparent cause, in my lumbar region, through the right hip, down to the knee. This was unusual for two reasons: one, ordinarily I would know what activity caused the spasm and shooting nerve pains. Two, my left side was always much more frequently involved than my right side.

However, two days before Thanksgiving I was getting dressed and putting my socks on. I remarked to Myrna, my wife, "Do you know something? This doesn't hurt; putting my socks on doesn't hurt!"

Let me explain. For 15 years I had not been without pain for a day. All of the daily, mundane activities would cause me pain.

So when I noticed I had no pain while putting on my socks, I decided to test myself in situations that would normally cause a great deal of hurt. Still no problem.

To this day, I have not experienced one bit of suffering in any situation — from driving a car long distances, which had been murder — to putting on my socks.

Looking back, I can see the possible reason in delayed healing. Since that night in August, a lot had happened to me and my wife spiritually. We met with one of the Melodyland pastors and rededicated our lives to Christ and have started re-reading the Bible, praying, singing praises and discussing the Word.

Many things have been changing, not just the healing of a crippled spine. So, I reckon, what God did that night in August was to get a foot in the door to my soul; then, when I really wasn't thinking about it, He healed my body.

Melodyland has had many different cases of healing. A boy from Canada was cured of cerebral palsy; a Mexican lady was healed of tuberculosis and diagnosed so by her doctor; a 72-year-old usher, who had many stomach ulcers for years, was prayed for, returned to his doctor and declared healed; a man who had arteriosclerosis, angina and congestive heart failure had a miraculous recovery, which his doctor confirmed after tests. An 11-year-old girl had been suffering from a hiatal hernia and surgery was recommended. She was on a bland diet, and had no appetite for six weeks. After prayer, her appetite returned, and she ate tacos. A specialist at Children's Hospital verified her recovery.

More and more healing testimonies reach my office: cystic fibrosis, eye injuries, fractures, hip deterioration, Lupus disease, deafness, diabetes, epilepsy, cancer, back trouble, asthma, blood disorders, bone cancer, numbness in the arms and internal bleeding . . . even big toes.

Once during a service, I felt impressed of the Holy Spirit to announce, "God is healing two big toes right now." I was amused, but sensing this was of the Lord, I persisted. "I said, God is healing big toes. There are two in the audience whom God is touching now. If you'll come forward . . ."

Later I received a letter from one of these individuals:

> Last Thursday during the service I was one of the two people who came forward with a healing of the big toe. When you said "big toe" something completely went over me, then down.
>
> I had just complained the night before to my husband that I would have to see my doctor and have the nail removed. I'd had the problem for 25 years, which I received from an injury.
>
> As a result, I had developed a fungus infection. At times it would flare up and cause considerable pain.
>
> When I finally realized it could be me, I said, "My toe, my big toe!" The lady next to me said, "Is it yours or mine?" I said, "I'll soon find out," and began removing my shoe. You can bet it would hurt to do that.
>
> I pinched and pinched; no pain. I decided to step out in faith. I just couldn't understand; why me, why had He picked me when I wasn't even asking?

Why Isn't Everyone Healed?

As in the Gospels, so in the Acts, we never read of anyone asking for healing and being denied. *Why, then, isn't everyone healed today?*

Kathryn Kuhlman was often asked this question, and perhaps no one in our times was more used of the Lord in a healing ministry. Here's how she would answer:

> Whether or not one was healed is in the hands of God.
> At no time was it my responsibility. But I'm human, and
> you'll never know how I hurt on the inside when I see those
> who came in wheelchairs being pushed into the street again.
>
> You'll never know the ache, the suffering I feel, but the
> answer I must leave with God.
>
> I do not know why all are not healed physically, but all
> *can* be healed spiritually, and that's the greatest miracle any
> human being can know.[4]

While we cannot say why everyone is not cured, we do
know there are hindrances to faith that keep many from receiving. One is disobedience. When seeking healing for our
physical needs, there can be no compromise with the adversary
for our souls because he is the author of disease.

Obedience, on the other hand, puts the individual on believing ground. James said:

> The prayer offered in faith will restore the one who is sick,
> and the Lord will raise him up, and if he has committed sins,
> they will be forgiven him. Therefore, confess your sins to
> one another, and pray for one another, so that you may be
> healed. The effective prayer of a righteous man can accom-
> plish much.[5]

It is God's will that you be in good health "just as your
soul prospers."[6] "If I regard wickedness in my heart," David
said, "the Lord will not hear."[7] The union of our hearts with
God's will is an important spiritual law, and faith always implies obedience. This releases the "Spirit of Him who raised
Jesus from the dead" and thus "He who raised Christ Jesus

from the dead will also give life to your mortal bodies through
His Spirit who indwells you."[8]

Many times people let lingering symptoms discourage
them or discount many healings. Often when they act in faith,
symptoms do not always disappear instantly. After Hezekiah
was healed, it was three days before he was strong enough to
go up to the House of the Lord. Mike Esses, Dr. Harrison and
Marvin Ford remained in the hospital for days after their experiences. Symptoms lingered for nearly three months in Dick
Humphrey's healing. Similarly I recall the miraculous healing of
Paula Blomgren, who suffered painfully from a disease called
escoliosis, which caused her backbone to curve in an S shape.
At 14 Paula was told she would have to wear a full body brace
for at least four years, perhaps much longer. In desperation,
she attended a healing service at Melodyland and later met
with four teenagers. Here's how she describes what happened:

> They talked to me for a minute and told me that they had
> been healed. I felt that God hated me, and this is why He
> put the disease on me.

> But after reassuring me that God does love me and that
> He was not responsible for my illness, they laid hands on my
> head. "Oh, Jesus, heal her," they prayed.

> Suddenly I felt something very warm inside my spine, and
> the pain seemed to go softly. I ran upstairs and took the
> brace off.

> I looked and my hump wasn't as large, and my shoulder
> wasn't down as much after the brace was off. I was beginning
> to look normal. So I went without my brace for two months
> and went to the doctor. I said, "Doctor, I'm healed. Jesus
> healed me."

> He took X-rays and said, "I don't understand this, but
> something has happened." It was a *gradual* healing. In two

months there was more difference than in the year and a half I'd worn the brace.

I haven't worn it for seven years, and I am straight. I feel so healthy, and I'm so grateful because I realize that God loves me.

Paula was associated with the Melodyland Drug Prevention Program, perhaps the world's largest hotline dealing with more than 100,000 calls a year. She later served in the Hotline choir.

The Bible differentiates between the gifts of healing and the working of miracles. Christ could do no miracles in Nazareth because of its unbelief. But He healed a few sick ones. If everyone was to be made perfectly whole instantly, it would all be miracles; there would be no place for the gifts of healing. Christ's promise is that all will recover, but He didn't say instantly. Symptoms of life remain in a tree for a time after it is cut down.

Faith means we are confident of what we've hoped for and convinced of what we do not see. Moffatt's translation of Hebrews 11:1 reads, ". . . convinced of course because God who cannot lie has spoken."

Sometimes people don't get healed or lose it afterwards because they're involved with some form of the occult. We teach people to renounce the devil and all works of darkness.

If someone comes to my office to be prayed for, I ask him to listen to a tape on Christian healing, which gives him the foundation of the Word of God before we pray. I think individuals can be healed without listening to the tape, but they will not be able to retain the healing as easily.

The tape is simply excerpts from the book, *Christ the Healer,* made with a professional radio announcer's voice and a musical background

Positive Confession

To be saved, the Bible says, we must confess with our

mouth and believe in our hearts that Jesus is Lord and is risen from the dead.[9] Is continued confession, therefore, necessary to obtain or retain healing?

There is an extreme teaching today that says we must continually confess God's Word to be healed.

The teaching is that the Word will heal *if* one continually confesses it. "God will make your body obey your confession of His Word," they say, "for no Word of God is void of power. God will make good all that you confess. Say it in your heart first, then aloud in your room. Say it over and over again; say it until your spirit and your words agree with each other. Say it until your whole being swings into harmony with the Word of God. You have exactly the same reason for expecting to be healed that you had for expecting to be saved. You have His Word for it. If you cannot accept that to the point of acting upon it, then your faith is still very far from what it should be."

While there is some truth in this teaching, it borders on dishonesty. If you still have a headache, admit it. You may say, "While I have not seen the evidence yet, I'm trusting God."

I have dealt with several cases where people claimed healing when they were not healed. This is presumption, not faith; it can bring a lot of harm to the body of Christ. If you are healed, you can have it checked out; any healing can stand the scrutiny of a doctor. You're *not* healed unless you have the evidence.

A tragic example of this extreme teaching on confession was a wheelchair case where a young man was told to demonstrate his faith. In attempting to get out of the wheelchair, he fell and broke his leg. The worst problem wasn't the broken leg, but his broken faith.

Another example of this is a local salesman who decided to throw away his glasses and claim healing. The man had difficulty seeing to drive. He had not received healing but was trying to force his faith. If a doctor has prescribed glasses, let him remove them.

Why God Heals Sinners

The question has been asked, *Why does God sometimes heal sinners instead of Christians?* Often after a miracle service this weighs heavily on me.

All of us have seen those with no church background receive their healing, while a Christian who has served the Lord faithfully did not. Perhaps one explanation is that it is often easier for a sinner to look totally to the Lord, and God performs the miracle—which again takes it out of the natural into the supernatural sovereignty of God.

Another observation: in Melodyland services there are far more conversions when miracles accompany the preaching of the Word. When a miracle of healing occurs, faith for forgiveness is inspired.

May I suggest that a sinner receives healing solely on the basis of grace, just as he does salvation? Of course, all healings are graces, whether for a sinner or a Christian. But there's only one prayer that Jesus obligates Himself to answer from a sinner and that's for forgiveness of sin. The believer, on the other hand, has become a citizen of a heavenly country. Thus, he has certain inheritance rights. In fact, the Bible declares that healing is "the children's bread," which means we can have divine health as one of the believers' benefits.

Why Do Men and Women of Miracles Die?

Kathryn Kuhlman's death came as a shock to the Christian world. Many have wondered how such a woman of faith, with a worldwide ministry of healing, could get sick, and why men and women of miracles die.

During one of the times when Kathryn was seriously ill, I made a trip to Tulsa, Oklahoma. At the same time, she and Oral Roberts were both hospitalized in separate medical cen-

Kathryn Kuhlman and Dr. Ralph Wilkerson on the Melodyland platform. Miss Kuhlman said she came to Los Angeles at Dr. Wilkerson's invitation and the leading of God.

ters. The newspapers carried the story. Oral Roberts was in for minor surgery on an old football injury, but Miss Kuhlman was in because of a heart problem.

For some time before her death, Miss Kuhlman was aware she had this problem. And on occasions when I was backstage with her she would whisper, "Please pray that my heart will hold out today."

While her ministry was powerfully anointed, she was a human being with a mortal body like all of us. Many people were shocked by her death because they thought her to be a timeless instrument of God. Not only did God grant many years of active ministry, but perhaps she compressed more into those years than any other person I've known.

Many questions have been raised about Miss Kuhlman's death. Some have suggested that God removed her because she resisted the extreme teaching on discipleship.

If this were true, other teachers correcting false doctrines would not survive. I consider this a dangerous judgmental position.

Others said she abused her body through overwork and lack of rest. There is no question that she kept an inhuman schedule, but her motive was to finish her work before the coming of the Lord. Often at the Shrine Auditorium in Los Angeles I heard her say, "No company could pay me enough to work the hours that I do." God could have empowered Miss Kuhlman with a supernatural strength to complete the task, but it was her time.

Those close to Kathryn have intimated that she died of a broken heart because of the deep wounds by a trusted friend. While this might have been a contributing factor, it is known that previously her heart was enlarged three times the normal size.

Though I realize that Kathryn was an earthen vessel and had imperfections like all of us, it is unquestionable that she was God's anointed. Once I was sitting beside a preacher at the Shrine during one of her services when Kathryn dramatically raised her hand and whispered into the microphone, "I'm just a handmaiden of the Lord." The preacher nudged me and quipped, "I only wish He'd called me to be a handmaiden!" In conclusion, it is evident that Kathryn Kuhlman's divine time had come and she kept her appointment with God.

When I was pastoring a new congregation in Topeka, Kansas, a young evangelist died whom God had used to pray for many people and who had tremendous miracles. Some of the congregation called me on the phone and said, "Pastor, we don't understand this."

The Lord gave me the Bible verse in Acts about Stephen. He preached the Gospel with such power and experienced all kinds of miracles in his ministry. Yet, in the prime of his work, he was stoned to death. We're sometimes tempted to say, "What a waste!" But it wasn't, because Stephen's martyrdom was probably what triggered the conversion of Saul of Tarsus, who stood by holding the cloak of one who threw stones. The ministry of Stephen was far greater through the life of Saul, who later became the Apostle Paul.

We can ask the same question about Jesus. Why did God take Him at such a young age—only 33? The whole world still needed to be reached. But when Jesus ascended, He sent the Holy Spirit and through the early Christians, the Gospel was spread throughout the world.

Jesus' death did not end the matter; it gave the Gospel a chance to multiply—just like a seed that's planted in the ground. When the seed is placed in the soil, it dies but in due time there is a resurrection, and life springs forth as a new plant. Death had to come before that seed could multiply.

So when any life that is committed to God is taken from this world, there is a multiplication. The individual who passes on becomes like a seed, and when God takes home His men and women of miracles it is to reproduce their work in greater measure.

Resurrection Present in Healing

There is no healing virtue outside the resurrection power of Jesus Christ, for every healing is a minor resurrection. The Bible says that the same Spirit, which had raised Jesus from the dead, has also quickened our mortal bodies. And the greatest power that was ever exhibited in the world was when Jesus came back to life and out of the tomb.

Death is pushing sickness to its farthest point. Just as resurrection is the answer to death, so healing is the answer to sickness.

The illness we suffer may not lead to death, but the forces of death are present. From the moment of birth, there are signs of decay, which progress as one gets older.

Those processes are reversed, however, the moment the resurrection power of Christ takes hold, giving us health. I've already said it's God's will that we be healthy. But there is a connection between spiritual life and physical prosperity. John wrote:

> I wish above all things that you be in health, even as your soul prospers.[10]

Now, this doesn't say Christians won't get sick, but that God wants people to be healthy and that spiritual laws have a beneficial effect upon the physical—if they are obeyed.

Some believe they are sick because they've sinned or are unworthy of God's blessings. The marks of sin do take their toll on the body, but Jesus forgives. He doesn't punish with disease, though illness can result from not taking proper care of the body.

Take an alcoholic, for example, who has cirrhosis of the liver. He comes to Christ and is forgiven, but that diseased liver is still there. It has to be healed by God's power.

While some suffer illness because of sin, this is not always the case. In John 8 Jesus was asked about a man who was blind from birth, "Who sinned? Was it he or his parents?"

Jesus said, "Neither."

We can never judge a case of sickness on the basis of sin. I have talked with some of the greatest saints of God who were ill; others had prayed for them, but they did not recover.

Many people who have had healing ministries are out of service today because of this kind of pressure. Often they themselves had been attacked physically, so they said, "If I don't have faith for my own needs, how can I pray for someone else?" A few have quit in frustration because they had prayed for some who got worse.

I must confess that many of the people to whom I minister are not healed.

We ought to be open about such cases and say, "Look, if someone gets healed, it's God. If someone dies, that's still God."

In healing it really is not our faith that operates; it is the Lord's. So if God gets all the glory, He also will accept all the responsibility. Leave the results to Him.

My son-in-law is another example of God's resurrection power in healing. At the time the accident occurred to this six-foot-nine athlete, he hardly knew me.

"Lord, I thank you because someone here is receiving a leg healing . . . I think it's a broken leg," I prayed as the Sunday night service was coming to a close at Melodyland. "I give you praise for the total healing of that leg."

A college student on crutches made his way outside. I did not see him during the service and barely knew him, for that matter.

Mike Dabney, who married my daughter, Debi, recalls:

> In the summer after the end of my senior year in high school, I was waterskiing with several of my friends. The mood was hilarious as it was more of a celebration. Most of us were in sports and had received scholarships for college.
>
> This was a last get-together before we split for different parts of the country. Clowning around, I climbed a tree to take a high jump into what I thought was deep water. Instead, I landed on a tree stump and busted my leg. I almost drowned, and if my friends had not been in sports and trained in how to handle a person with a broken leg, I could have had permanent damage.
>
> At the hospital the doctor pieced the leg back together, putting 12 inches of steel in it.

"Your leg was a shattered mess. You may never be able to walk on it again, and if so, you will have to use a cane," he informed me soberly.

I was in a cast extending from the bottom of my toes to my hip, and the doctor continued, "You will be in the cast for a minimum of two years."

After I went home, it was very boring just sitting around all the time. I was beginning to feel sorry for myself that day when the telephone rang.

"Hello."

"I'd like to speak to Mike Dabney."

"This is Mike."

The coach of a college in Southern California asked me to come out and be interviewed for a basketball scholarship.

"Uh . . . Sir . . . I have a broken leg."

"I've heard about that, but I still want you to fly to California so we can talk together," the coach persisted.

Hanging up the telephone, I thought, "Guy, if this man is stupid enough to want to talk to me about a scholarship to play basketball, even knowing I have a broken leg, then I am going to be smart enough to go out there."

When I got to California and sat down to talk with the coach, he asked me how long the doctor thought I would be on the crutches. I told him the whole truth. I couldn't even take the body cast off for at least 24 months, and the doctor said I will never walk without a cane.

To my surprise the coach replied, "But I believe in miracles — I think you will be playing basketball again maybe in a semester or two. I'm still going to give you the scholarship."

On Sunday some friends of mine took me to Melodyland. As always, being a preacher's kid, I sat on the back row as far away as I could get.

The pastor, whom I didn't know at that time, asked us to stand to be dismissed. When the pastor said, "Lord, I thank you because someone here is receiving a leg healing . . ." at that moment I felt all the muscles and the bones in my leg become strong. It was as though my leg was trying to break out of the cast, just as Jesus burst out of the grave. I knew I was healed.

The next day, I flew back to Texas and went to my doctor telling him to X-ray my leg. I was sure it was all right now. "The X-rays were fine," the doctor reported. But he refused to remove the cast.

I packed and returned to California two days later. I knew that my leg was healed, so I went to another doctor. After making three sets of X-rays, the doctor said, "There's absolutely nothing wrong with that leg." Immediately he removed the cast.

After the muscles rebuilt, I was working with the team before the first semester ended and continued playing basketball on a scholarship through college.

Medical Profession and Healing

How does the medical profession view divine healing? Many physicians are skeptical, of course, but they cannot deny the miraculous. Dr. Elizabeth Kübler-Ross speaks effectively on this in suggesting a role for attending physicians where a terminal patient insists he is healed.

If a young man with a terminal cancer makes a statement that he is miraculously healed, this means to me he wants to believe in a miracle in spite of the fact that from a medical point of view he's regarded as terminally ill.

I would sit with him and say, "Yes, miracles do sometimes happen;" and wait for a while and continue to visit with him so that he has an opportunity to share with you his feelings about his terminal illness or his belief that he has been cured.

It is not your role, whether you are a member of a helping profession or a family member, to break down a defense . . . If you do not believe, yourself, that miracles do happen sometimes, you can simply ask him to tell you more about it. He may even end up convincing you.

Over the last eight years we have had several patients who had been given up and from a medical point of view had no practical chance of recovery, but who are *still alive* several years after the predicted date of their death.[11]

Our Creator longs for His children to fulfill their earthly missions and return to Him. From the beginning, He knows our roles in life and how much time our pilgrimage requires. If one dies and is raised up, divine intention is implied. Extension can have only one premise—to bring even more glory to the kingdom of God.

I believe in healings and resurrections from the dead because I have witnessed them. But my greatest reason for believing is based on the Word of God. There are cases of resurrection in both the Old and the New Testament.

In the next chapter we will examine them and tell why beyond and back experiences were significant in the early church.

. . . they of the people and kindreds and tongues and nations shall see their dead bodies three days and an half, and shall not suffer their dead bodies to be put in graves (Rev. 11:9).

9
You Will Be Raised From the Dead!

Someday, maybe within the next 10 years, your favorite television show will be interrupted with the following news bulletin:

"Good evening, ladies and gentlemen. We interrupt this telecast to bring you some incredible news. The two men who have been causing the world so much trouble are now being confronted by our world leader.

"Now, we take you by satellite[1] to a street in Jerusalem where the Leader is doing what no man has dared to do for the last three and a half years!"

The screen goes blank for a second. Then a reporter with a microphone appears, talking excitedly. "Look at it, ladies and gentlemen! Our Leader, the world's Antichrist, is actually challenging these prophets. Would you believe it? The troublemakers are having difficulty. Never has this happened! What a mighty ruler this world now has. I can't imagine . . . these two men are losing.

"Wait! They've just been struck down. We'll get the camera in closer. I believe they are dead."

The Antichrist's press secretary comes into view, and the newsman interviews him.

"Yes, a doctor is examining them now to see if they're dead," the press secretary says jubilantly. "This is the moment the whole world has been waiting for. The ruler decided it was time to have a showdown with these troublemakers. What an amazing man he. . . ."

The newscaster interrupts.

"It is so. They are dead. The doctor's given his word. Our great ruler! Who is able to make war with him?"

The camera comes in for a close shot of the dead prophets.

"Here they are, people of the world," the newscaster says excitedly. "Here are the men who caused us so much misery. Now they are finished. These strange men have been preaching for three and a half years on the same subject: judgment.[2] Now they've been judged.

"All of us believe that man is getting better all the time and that everything is relevant. These men have said this is a lie.

"But the most blasphemous statement these self-proclaimed prophets have made is that the noted imposter Jesus Christ is now Lord of all.[3]

"Their sermons have been backed up by fanatic happenings, such as fire breathing out of their mouths if anyone merely thinks wrong about them.[4] Severe drought has occurred, for which these men take credit. They've also turned many of the world's streams into blood. Attempted assassinations have been frequent, but with no success.

"The whole world has been looking for a strong leader, someone with not only the ability to plan proper strategy, but who has supernatural flair. At last we have our leader, and he has been able to overcome these troublemakers."[5]

As the newscaster continues his special report, someone approaches and hands him a slip of paper.

"Ladies and gentlemen, I've just been informed . . . yes, it's true . . . there already are reports of some rain falling in parts where there has been no precipitation for three and a half years.

"And here's something else, ladies and gentlemen! The streams are running clear with fresh water again. The blood disappeared as suddenly as it came."

Turning again to the press secretary who had just approached from talking with the director, the newscaster asked, "Sir, what will happen to their bodies?"

"They will lie in the city streets for three and a half days, a day for each of the years they've tormented us. No one will be allowed to move or touch them.[6]

"We suggest you send gifts[7] to your friends," the press secretary continued. "This will materially stimulate our suffering commerce. Trade will improve. It will give the shops an incentive to open up again to full production."

"It's almost like Christmas," the newscaster joined in, as once again the camera flashed to the death scene. "In fact, it is better, because we have eliminated the foolish aspects of Christianity in this special holiday."

"We should call this the Dictator Day. It's only appropriate that you send choice gifts to the ruler who has delivered you from a great menace," the press secretary added. "And don't forget your local officers. This is an order! This celebration is to last one week. . . ."

The screen flashes back to the local television station, where the announcer closes, "We return you now to our regular programming. We will issue other bulletins as soon as they become available."

The next day a news special has been prepared, and the newscaster is presenting his report:

". . . it seems unbelievable that a little country of only eight miles wide and 33 miles long continues to be the center of world news. Only 7,992 square miles in Palestine, but no part of the world has shown such remarkable economic growth. Perhaps the reason is because of the great mineral deposits of this part of the world and the drastic needs around the globe.

"Of course, it shows human ingenuity, too, particularly when we no longer have to tolerate Christianity. Certainly our people are religious, but now Christianity is finally stamped out."

An image of the temple area in Jerusalem appears on the screen as the announcer's voice continues:

"See, the temple area is now functioning as it was many years ago.[8] These so-called Christians have vanished off the scene, making room now for a more intelligent approach to life.

"The two weird-looking characters, whom our world leader killed yesterday, called themselves prophets of God. Their dress of sackcloth and manner of life suggested that they indeed had strange powers. Some nicknamed them Moses, Elijah or Enoch because of the similarity in their mode of living."

Again the cameras capture the bodies of the two prophets as they lay on the street.

"Here they are," the announcer continues, "the two witnesses, who called themselves prophets of God. Evidently their God is dead," he says mockingly.

The scene changes to the airport.

"Excursion planes are arriving here from many parts of the world for the thrill of personally seeing these two madmen. Under the leadership of our dictator, conditions will return to normal very soon."

The cameras take viewers to various post offices, where large volumes of mail are being sorted for delivery.

"Already the mails are packed with greetings and gifts, as suggested yesterday by the world ruler," the newscaster continues. "Business is better than it has been for months. The city of Jerusalem has witnessed more merriment than it has for years. Someone has found a Bible and says much of these conditions are described in Revelation 11. However, none of us intelligent beings believes this book anymore. . . ."

The middle of the next afternoon . . . television programming is interrupted on all channels throughout the world.

"Ladies and gentlemen, we hope we can get this newscast through. Another catastrophe has struck. In Jerusalem an earsplitting crash of thunder has just sounded. The two prophets have suddenly stood up.[9] They're alive! Incredible! Just incredible, ladies and gentlemen!"

The screen flickers as satellite communications capture the event as it happens. "I can't believe it," the newscaster shouts into the microphone. "Look at them . . . they are rising. They are going up without the aid of an aircraft. Now they are disappearing into that cloud. They're gone. . . ."

Suddenly the earth begins to rumble and shake. The cameraman loses his balance. Communications are lost for a second, then they're back on. The newscaster breathlessly describes what he sees: "Large cracks are appearing in the sidewalks and streets, ladies and gentlemen. The quake is getting harder . . . walls are collapsing, people are being buried under debris.

"I hope you are still receiving this telecast. A major earthquake is rocking the city . . . damage will run into the millions. There has never been an earthquake recorded in Jerusalem with such magnitude . . . many of the people have lapsed back to their traditions and are standing in the streets, crying to their God for mercy.

"A second tremor has just been felt. We are. . . ."

The cameraman is struck by falling debris, and the picture goes blank. . . .

"Ladies and gentlemen, please stand by. Please stand by. We're experiencing technical difficulties. . . ."

Suddenly, the rumbling and shaking in Jerusalem grow weaker, then stop. The dust begins to settle as the cries of the injured and the survivors express their terror and grief. Many newsmen at the scene are among the victims. Those who survive try to capture the story while the media around the world rushes correspondents to the scene.

Soon the story emerges. One-tenth of the city is in rubble; seven thousand men, plus women and children are dead. . . .[10]

*   *   *   *

Yes, this very city of Jerusalem will witness the most spectacular beyond and back experience of modern time since the resurrection of the righteous at the return of Christ for His church.

As I write this chapter, the sun is just coming up over the hills of Jerusalem. Today will be my day to stay behind to write, meditate and pray while the others go on tour. I have always been fascinated by the truth that Biblical incidents actually happened here. And my mind is racing back in time to other resurrections, some of which took place in Old Testament days.

Elijah
I Kings 17:17-24

The rugged prophet Elijah lived in the midst of miracles. On one occasion, in 63 words and three minutes, he called down fire from Heaven and confounded the prophets of Baal.

What a strange paradox that death would strike the son of a widow of Zidon who had seen God provide food miraculously during famine because of Elijah's faith.

She had only enough food in the house to cook a "last meal" for herself and the boy. Then Elijah came along and said, "Bake me a little loaf of bread first," promising that God would multiply the flour and oil in her containers until the rains came and the crops grew again.

The widow obeyed, and God kept His word. Then tragedy struck. She recognized Elijah as a man of God who could intervene. She believed that the lad's death was a mark of divine displeasure. But the Bible doesn't teach this. "God has not dealt with us according to our sins, but rather according to his mercy."

Elijah took the boy upstairs to the guest room where he lived, laid him on the bed and began to pray.

"Oh Lord, my God," he cried, "why have you killed the son of this widow with whom I am staying?"

Elijah stretched himself out on top of the boy three times, the Bible says, and cried out to the Lord.

There was resurrection power and faith in that prophet as he said, "Lord, let this child's spirit return to him!" And God heard it. He always hears the prayer of faith. The child's spirit re-entered his cold body. Suddenly, he began to breathe; his eyes opened. His temperature became normal.

Elijah helped the boy sit up, then stand. Imagine the excitement Elijah felt. He was used to miracles, but always they amazed him. Elijah grabbed the boy's hand, and they bounded down the stairs.

"See! Your son is alive!" he beamed.

"Now I know you're a man of God," the widow cried jubilantly. "And I know that whatever you say is true and from the Lord."

Legend says the raised lad became a servant of Elijah and later of the prophet Jonah.

Elisha and the Shunammite
II Kings 4:18-37

The son of a wealthy Shunammite family had a sunstroke one day while he was with his father in the fields. The father instructed a servant to take the boy home to his mother, where the lad died a short time later.

The mother, who a few years earlier had witnessed a miracle at the word of Elisha, summoned the prophet. When he entered the room where the child lay, he shut the door behind him and prayed.

"Lord, this mother needs her son. She needs a miracle. Send this lad's spirit back into his body."

Suddenly, Elisha felt prompted to lie on top of the child. He placed his mouth on the child's mouth, his eyes on the child's eyes and his hands on the child's hands. Soon, the body began to feel warm.

He stretched himself out again on the boy. This time the lad sneezed seven times and opened his eyes.

So the Shunammite woman received a double miracle. God had removed her barrenness years before. Now He had raised her son from the dead.

Elisha and the Dead Moabite
II Kings 13:20, 21

Elisha's last miracle was a post-mortem experience.

Several years after the resurrection in Shunam, the prophet died and was buried, the Bible says.

Bandit gangs of Moabites would invade the land each spring in those days. One day some men were burying a friend near Elisha's tomb when they spotted the marauders.

"Quickly, let's toss his body into this tomb," one said. "We'll come back later and bury him. We must get out of here. Hurry! Before the Moabites see us!"

One grabbed the dead man's feet, while the other took hold of the arms. Then they threw him into Elisha's tomb.

As soon as the body touched Elisha's bones, the dead man revived and jumped to his feet.

Tradition says that the unknown man who was resurrected lived only an hour. God wanted all generations forever to remember that His power is not limited in death, for in Him, in Christ, is the resurrection and the life.

Jonah
Jonah 1:1-3:10

Jonah, often considered the greatest preacher of the Old Testament, was sent by God to go to Nineveh.

But Jonah didn't want to go. He ran from God. The way from God is always downward. Jonah went down to the sea, found a ship going down to Tarshish, and went down into the hold of the boat to sleep. Later he was thrown overboard and went down to the bottom of the sea, where he was taken into the belly of the fish.

Never had there been a more fervent prayer meeting, nor a more repentant soul than Jonah's was at that moment.

As the waters had closed above him and the seaweed wrapped itself around his head, he'd found himself in the stomach of the fish. He gagged at the stench and gasped for air as he tumbled around inside.

"Oh, Lord, you've rejected me," he cried. "You've cast me aside. I've been expelled from your sight!

"God, give me another chance! How can I thank you enough for all you've done for me in the past. If you'll give me another chance, I'll surely fulfill my promise to you!"

The Bible says the Word of the Lord came a second time to Jonah. How often God has given each one of us a second chance.

For three days and nights Jonah bunked in the belly of the fish. The Bible isn't clear whether Jonah lived or died, but there are several reasons to believe he died.

First, without oxygen and trying to breathe in the fish's digestive fluids, only a miracle from God could keep him alive, even for a few minutes.

Second, Jonah said, "I cried for help from the depth of Sheol. . . . Thou has brought up my life from the pit." This is an indication that Jonah's spirit had departed from his body and had gone to the place of departed spirits.

Third, it would be perfectly reasonable to assume that Jonah died because Jesus used this incident as a type of his own death and resurrection. If you question that Jonah died, would it be possible to question that Jesus died?

> As Jonas was three days and three nights in the whale's belly; so shall the Son of man be three days and three nights in the heart of the earth.[11]

Jesus uses the word *cathose,* meaning "even as," in making the comparison between His death and resurrection and Jonah's stay in the fish. *Even as* is the strongest Greek comparative He could use, so we can expect the maximum number of points of comparison.

Since Jesus couldn't have made a stronger analogy, we can look for every point as similar.

When God finished working Jonah over, He ordered the fish to spit the prophet up on the beach.

Weak from his ordeal and covered with debris from the fish's stomach, Jonah crawled slowly through the shallow water, rinsing himself as he made for the shore. It was early morning and the sun and breeze felt marvelous.

"Jonah," a Voice called. "Jonah, get up and go to Nineveh. Warn them about the judgment I'm sending, as I told you before!"

Jonah picked himself off the sand without saying a word and began his journey, stopping only for brief periods of rest and food.

"Forty days from now Nineveh will be destroyed!" the prophet shouted to the crowds that gathered about him on the first day he entered the city.

If all preachers would preach what God told them to preach, the most wicked city could be won to Christ. As a result of Jonah's eight-word sermon, this city of about 175,000 persons repented.

Many other incidents of resurrections are mentioned in the New Testament.

Jairus' Daughter
Mark 5:22-42

A miracle within a miracle is about to occur. Jesus had just healed a woman with an issue of blood, which she'd had for 12 years. At the time her infirmities started, a girl was born to Jairus, a leader of a synagogue in Capernaum. . . .

There was a stir and mumblings at the edge of the crowd where Jairus had been standing. A messenger was whispering something in his ear. Suddenly, Jairus began to weep.

"It's too late, Jesus," the messenger said sadly. "Jairus' daughter just died; there's no point in your coming now."

Jesus walked over to Jairus. "It's never too late. Don't be afraid, Jairus. Just trust me."

He turned toward the crowd. "I must go on with Jairus and my disciples alone," he said, then followed Jairus and his messengers, who had already started ahead.

The afternoon sun was still bright and warm, and Jesus shielded His eyes as He and Peter, James and John quickened their pace to catch up with Jairus.

When Jesus arrived, mourners were unrestrained in their weeping and wailing. He walked inside Jairus' fashionable home.

"What's all this weeping and commotion about?" He asked, dismayed at the confusion. "The child isn't dead . . . she's just asleep!"

There was stunned silence for a moment, then bitter laughter. "Asleep, He says!" they jeered. "The child's dead!"

"All right," Jesus said firmly. "Everybody out. Out!"

There's a communion in the Spirit that requires privacy, except for believers. Jesus told Jairus to bring his wife, then beckoned to His disciples: "Peter . . . James . . . John, come with us."

They went into the bedroom where the girl lay on a bed. Jesus walked over to the body. It was cold, and the first signs of rigor mortis were setting in. He looked at her tenderly for a moment, then bent down and took her by the hand.

"Little girl, arise!"

Immediately her spirit re-entered her body,[12] and she got up and began walking around the room. A few minutes later the child was given something to eat.

Widow's Son at Nain
Luke 7:11-17

Many of those who had witnessed His miracles in Capernaum followed Jesus to Nain, a village in Galilee about 18 miles away. Some from Nain who had heard Jesus was coming came out to meet Him.

As usual, it was hot, and Jesus was weary from the long, dusty walk. He was anxious to reach town, rest and have something to eat before He would again minister to the people. As Jesus approached the gate, a funeral procession emerged. Jesus stopped and leaned on His staff for a moment, surveying the scene.

"The dead one is a boy, and he was the only son of a widow," one in the crowd from Nain who was standing near Jesus observed.

Jesus watched as the mourners filed by, but when the widow passed, His heart overflowed with compassion. Jesus approached her. "Don't cry," He said softly, then walked over to the coffin and touched it. With that, the bearers stopped. Jesus' reputation for miracles was widely known, and the crowd watched in curious excitement, wondering what Jesus was going to do.

"Young man, arise!" Jesus commanded.

Immediately, the boy sat up and began to talk. Jesus

helped the lad out of the coffin and presented him to his mother as the crowd gasped in amazement and praise to God.

"A mighty prophet has risen among us," some cried.

"We have seen the hand of God at work today," others joined. Soon the news of the boy's resurrection spread throughout Judea and across other borders.

Lazarus
John 11:1-45

Bethany was Jesus' favorite resting place, and it was the home of Mary, Martha and Lazarus, who were close friends of His. It was in Bethany, about a month before His own death and resurrection, that Jesus performed the last of seven miracles recorded by John.

Jesus was near the Jordan River where John the Baptist had first baptized. Many followed Him there, and He was teaching and healing the sick.

Meanwhile, Lazarus had taken ill in Bethany, and his sisters sent word to Jesus.

Jesus stayed where He was for a couple more days and continued to minister to the people.

By the time Jesus and His disciples arrived in Bethany, Lazarus had been in the tomb four days. Since Bethany was only a couple of miles from Jerusalem, many of the Jewish leaders had come to pay their respects and to comfort Martha and Mary.

When she heard Jesus was coming, Martha went down the road to meet Him before He reached the village.

"If you'd only been here," she said, trying to control her tears, "Lazarus wouldn't have died. But it's never too late for you. God can bring him back to life if you ask him."

Jesus was silent for a moment as He looked intently at her. "He'll come back to life again," Jesus said finally.

"Yes, I know. When everyone else does on resurrection day," she broke out.

Jesus comforted her as they started walking. His voice was tender. "Martha, I am the resurrection and the life: he who believes in Me shall live even if he dies."

Martha stopped and looked at Him questioningly.

"And everyone who lives and believes in me shall never die. Do you believe this?"

"Yes, Lord; I have believed that you are the Christ, the Son of God, even He who comes into the world."

Jesus and His company had not reached Bethany when Martha found them. Jesus decided to pause for a while and rest. Martha took advantage of this and rushed back to town for Mary.

At home, Martha called Mary aside from the other mourners and whispered, "Jesus is just outside town, and He wants to see you."

Mary left quickly and followed Martha to meet Jesus. Some of the Jewish leaders who were there assumed she was going to the tomb to weep, so they followed also.

"If you'd only been here, my brother wouldn't have died," Mary cried bitterly when she saw Jesus.

When He saw her tears and the weeping of the mourners who accompanied her, Jesus was deeply moved. "Where is he buried?" Jesus asked.

"Come and see."

Tears stung Jesus' eyes, welled, then rolled down His cheeks and disappeared into His beard. It was the first time anyone had seen Him weep.

"They were close friends," some of the mourners said among themselves. "See how much He loved Lazarus?"

"Maybe . . . but He healed a blind man. Why couldn't He keep Lazarus alive?" another sneered.

Jesus overheard the comment, and His anger at their cynicism and unbelief began to boil.

The tomb was a cave with a heavy stone rolled across the entrance.

"Remove the stone!" Jesus commanded.

"But Jesus, he's been dead four days, and the stink is horrible!" Martha objected.

"Didn't I tell you, Martha, that if you believe, you will see the glory of God?"

Jesus gestured to the men to remove the stone, then looked into the sky. "Father, I thank you because you've heard me," Jesus prayed. "I know you always hear me. But because of these people around me, I have said it so they will believe that you sent me. . . ."

When Jesus finished praying, He lowered His head and looked toward the open tomb. He took a deep breath.

"Lazarus," He shouted. "Come forth!"

The crowd waited breathlessly. Some ready for the greatest miracle they had ever seen Jesus perform. Others . . . waiting for another cause to ridicule.

Suddenly the figure of Lazarus appeared at the entrance of the tomb, still wrapped like a mummy in the gravecloth, his face muffled in a head cloth.

Jesus broke the silence. "Unwrap him and let him go."

Because of this miracle, many of the skeptics among the Jewish leaders believed. Others, however, quickly left to report to the Pharisees.

Jesus stopped His public ministry among the Jews after this incident and went to Ephraim, a village on the edge of the desert.

Dorcas
Acts 9:36-42

Dorcas was like a clock: open face, moving hands and full of good works. In the ancient seaport of Joppa, she was known for her concern for widows and the poor. Her occupation was making underclothes and outer garments.

One day she became ill and died. Friends prepared her for burial and placed her in an upstairs room. When they learned that the Apostle Peter was in nearby Lydda, they sent for him. He came immediately.

When Peter arrived, he was escorted up the stairs to the room where Dorcas was lying. The room was filled with weeping widows, dressed in mourning clothes, showing each other the garments Dorcas had made for them.

After a brief conversation with those who had escorted him to the room, Peter's big fisherman frame seemed to stand taller with commanding authority. He began to feel the Spirit of the Lord upon him.

"Ladies," he announced, "I'm sorry, but I have to ask you to leave this room."

Quickly and silently, the women filed out the door as Peter knelt and began to pray. No doubt Peter remembered the same act of Jesus raising Jairus' daughter and felt it was wise to be alone with the Lord at such times as these. The anointing of the Holy Spirit swept through Peter as he turned toward the body, still on his knees.

"Dorcas, get up!" he commanded.

Her eyes fluttered slightly, then opened wide. Her breathing started, and her flesh began to look normal. When she saw Peter, she sat up.

"Here, Dorcas," Peter said happily, offering her his hand. "Let me help you up."

Dorcas stood to her feet, a bit shaky at first, then steady. Peter walked to the door and called down to the believers and the widows who were waiting expectantly for a miracle.

"Come on up," he cried. "She's alive and asking for you!"

Some of the women bounded up the stairs, while others raced through town with the news. Many believed in the Lord because of this resurrection.

Eutychus
Acts 20:1-12

The resurrection of Eutychus happened about midnight on

a Sunday. The believers in Troas had met for a Communion service, and because the Apostle Paul was due to leave the next morning, he had been asked to preach before they took of the Lord's Supper.

He had an unusually long sermon that night. Numerous lamps had burned out most of the oxygen in the crowded room, and Eutychus had perched himself on the window sill to escape the stuffy room and get some air. He was slightly bored, anyway. The boy tried his best to keep awake, but finally dozed off, lost his balance and fell three stories to his death.

"Eutychus fell out the window!" screamed his mother, running out the door and down the stairs. Paul and the others followed close behind.

The Apostle pressed his way through the group and took the boy from his mother's arms. The anointing of the Spirit was so heavy on Paul that when he embraced the youth, he came back to life.

"Don't worry," Paul announced to an amazed congregation. "He's all right. He's alive."

They began to praise the Lord right there on the street! Finally, they went back to the room and held the Communion service. Then Paul preached another long sermon. It was dawn before the meeting ended.

I have often wished that Paul would have interviewed Eutychus about his beyond and back experience, but the saints considered it more important to have Communion together.

St. Paul
Acts 14:20

Paul was well acquainted with the resurrection power of Christ, for he had experienced it himself sometime before the Eutychus incident.

Paul and his companion Barnabas had received wide acclaim for the healing of a man who had been lame from birth.

"These men are gods in human bodies!" the crowds said.

They took Barnabas for the Greek god Jupiter, and Paul for Mercury. The priest of the Temple of Jupiter on the outskirts of Lystra, where the miracle had occurred, brought Paul and Barnabas cartloads of flowers and prepared to sacrifice oxen to them before the crowds.

"Men, what are you doing?" Paul and Barnabas had cried as they ran among the people. "We're just human beings like yourselves. . . ."

Paul preached the Word to them. Yet he and Barnabas could barely restrain the crowd from worshiping them.

It was this same crowd, a few days later, which was swayed against the evangelists by enemies from Antioch and Iconium.

The crowd became so incited against them, they stoned Paul and dragged him outside the city, leaving him for dead.

But while fellow Christians were gathered about Paul, he got up and went into the city.

The question has often been raised whether Paul really died. I believe he did.

First, because his enemies believed he was dead. The Christians around his seemingly lifeless body certainly were helpless. A healing or a resurrection had to occur for Paul simply to get up and walk into the city, then immediately continue his missionary journeys.

Second, if this is the time when Paul died, then he is giving to us his beyond and back experience in II Corinthians 12.

"I knew a man in Christ about fourteen years ago," Paul wrote of himself. "(Whether in the body, I cannot tell; or whether out of the body, I cannot tell: God knoweth;) . . . how that he was caught up into paradise, and heard unspeakable words, which it is not lawful for a man to utter."[13]

Here Paul says he doesn't know whether he was dead or

not, but he was caught up into the third heaven.

Ask Paul what he saw, and he responds, "You may think I am gifted in expression, but this is impossible to describe. Also, I am forbidden to tell some of God's secrets."

Paul never took advantage of this publicity of being raised from the dead, nor of the marvelous revelation that God gave him.

Saints Rise from Their Tombs
Matt. 27:50-53

It had been dark since noon as Jesus hung in agony on the cross, and about three o'clock the pain of His loneliness had reached its peak.

"My God! My God!" Jesus shouted. "Why have you forsaken me?"

"He's calling for Elijah," someone jeered as he filled a sponge with sour wine, put it on a stick, and held it up for Jesus to drink.

"Aw, let Him alone," others pleaded, mockingly. "Let's see whether Elijah will come and save Him!"

The sound of laughter rang in the air, interrupted only by one final cry of agony as Jesus' spirit left His body, and He died.

Suddenly, the earth began to shake. Rocks split, and the earth opened up. The veil, which covered the holy of holies in the temple, was torn from top to bottom; tombs opened and many bodies of the saints who had died were raised.

This is one of the most startling accounts in the Bible of saints being resurrected. They were raised from the dead when Jesus died, but they did not leave their tombs until Christ's resurrection early that Sunday morning. Then they went into Jerusalem and appeared to many.

The purpose of this resurrection was to transfer the saints from their place of abode in "Abraham's bosom" to Heaven, where they now await the resurrection to judgment.[14]

Some suggest that those who came out of their graves at Christ's resurrection were Old Testament saints. But it is more conceivable to believe that they were those who had heard Jesus preach during the 42 months before His death. They were likely friends of Jesus, such as John the Baptist, Zechariah, Joseph, Anna and those who had acquaintances in Jerusalem.

Can you imagine how shocked the relatives of these resurrected people were when they came to their door?

They might have had glorified bodies and ascended later with the Lord. Or they could have been temporarily restored to their mortal bodies, only to die a natural death later. One thing we can be certain of: they came forth to witness their immortality and testify by their own experience that the Redeemer had won the victory over the grave.

Later, New Testament writers do not mention these resurrections. Nor any of the saints who returned from the grave. It was a momentary release, a rapid penetration. It was God's counteraction to unbelief. This event was an announcement to the saints. The news was bad at the time, for no one knew of Christ's resurrection. Christianity for a few days was stopped. These resurrections were added evidence to Christ's resurrection to get the good news out.

Resurrection to Come
I Thess. 4:16, 17

Today, we are looking forward to the greatest event of all since the resurrection of Christ. It is the day when our Lord returns from Heaven with a shout of triumph, "and the dead in Christ shall rise first: then we which are alive and remain shall be caught up together with them in the clouds, to meet the Lord in the air: and so shall we ever be with the Lord."

Students call this the general resurrection of the religious dead.

It will be a great celebration, a time when millions will get new bodies, immortal and incorruptible.

When Christ rose from the dead, He became the first of millions who will come back to life again some day, Paul says in I Corinthians 15.

Hal Lindsey spoke at Melodyland recently on the order of resurrection. Here are excerpts from his message:

> "Now Christ has been raised from the dead, the first fruits of those who are asleep. For since by a man came death; by a man also came the resurrection of the dead. For as in Adam all die, so also in Christ all shall be made alive. But *each in his own order*" (I Cor. 15:20-22).

> That's the thing I want to stress: "each in his own order." Christ is the "first fruit" in the resurrection order. After that, those who are Christ's at His coming.

> Now the words "first fruits" in the illustration here refer to the order of the Jewish harvest. First, when they would harvest the grain, the first fruits would be a token of the harvest; this would be brought before the Lord at the temple.

> They would offer it to the priest, and it would be called a wave offering. This would be considered a guarantee of a full harvest.

> Then came the general harvest, followed by the gleanings. The gleanings were the grains left by the cutters for the poor.

> Christ is the first fruit. The general harvest is going to be the rapture, the raising of the church and the Old Testament saints.

> Let's look at a verse on that. Jesus said in John 14:1-3, "Let not your heart be troubled; believe in God, believe also in Me. In My Father's house are many dwelling places; I go to prepare a place for you. And if I go and prepare a place

for you, I will come again, and receive you to Myself; that where I am, there you may be also."

He's preparing dwelling places for us, and He said He's coming to take us to those prepared places.

When Jesus comes back in the second advent, He is coming back to the Earth visibly and personally. It will be at the end of a seven-year period of global war, and He will put down all the armies that will be in battle around the Middle East.

At that time He doesn't take believers off the Earth back to His Father's house. He keeps them on the Earth to rule with Him in a kingdom that will last a thousand years.

But He says of the believer in the church that he will be caught up to be with Him in His Father's house, which is not on the Earth.

This tells me that these advents are not the same events. First, Christ is coming for His church to take us to His Father's house. Then He's coming back to the Earth in a visible return to set up a kingdom.

When Jesus comes to take us to the Father's house, this is the second stage of the resurrection. "The Lord Himself will descend from heaven with a shout . . . and the dead in Christ shall rise first. Then we who are alive and remain shall be caught up together with them in the clouds to meet the Lord in the air . . ." (I Thess. 4:16, 17).

Now what takes place after that? A period of tribulation follows; then there is another resurrection.

"And I saw the souls of those who had been beheaded because of the testimony of Jesus and because of the Word of God, and those who had not worshiped the beast or his image, and had not received the mark upon their forehead and upon their hand; and *they came to life* and reigned with Christ for a thousand years" (Rev. 20:4).

This is talking about the raising from the dead of the Tribulation saints. In the illustration of the harvest, this is the gleaning.

Now it says (verse 5) that the rest of the dead did not come to life until a thousand years were completed, calling this the first resurrection.

In other words, it's finished at that point. The first resurrection is the resurrection of life, and all the believers are raised in those stages. Christ the first fruit, the general harvest, the gleaning.

Who are the rest of the dead? Only unbelievers. Verse 6 says, "Blessed and holy is the one who has a part in the first resurrection; over these the second death has no power, but they will . . . reign with Him for a thousand years."

Jesus predicted the resurrection of the righteous.

"Don't be surprised at this," He told His disciples one day. "The hour is coming when the dead will hear my voice, and they that hear will live."[15]

Our resurrected bodies will be human, but immortal; without blood, able to walk through walls or doors and to travel through millions of light years of space in seconds. A study of the nature of Jesus' resurrected body reveals this, and the Bible says, "We will be like Him."

What was Christ's body like after His resurrection?

Before dawn on resurrection morning, Mary Magdalene came to the tomb and discovered that the stone had been rolled away. Frightened, she ran to tell Simon Peter and John.

Peter and John raced to the tomb to investigate. John got there first, stooped and looked in and saw the linen cloth there. Nearly out of breath when he arrived, Peter pushed his way past John and went inside. The face cloth that had wrapped Jesus' head was neatly rolled into a bundle, while His grave clothes lay just where His body had been. It appeared as though His body just passed through the grave clothes when Jesus arose.

The two disciples left for their homes, so overjoyed at the news they didn't notice Mary outside the tomb weeping.

She stooped and looked into the tomb, curious about what Peter and John might have seen. She was startled. Two white-robed angels were sitting at the foot and the head of the place where Christ's body had been lying.

"Why are you crying?" they asked.

Her lips quivered as she tried to choke back the tears. "Because someone's taken away my Lord."

She caught a movement out of the corner of her eye and glanced over her shoulder. Jesus was standing behind her, but she didn't recognize Him.

"Why are you crying? Who are you looking for?" He asked.

Thinking he was a gardener, she said, "Sir, if you've taken Jesus away, please tell me where He is so I can find Him."

"Mary!" The man's voice suddenly sounded familiar.

Mary whirled around. "Jesus!" she exclaimed, reaching out to Him.

"Don't touch me!" Jesus stepped back. "I haven't as-cended to my Father yet. But go find my brothers and tell them that I am ascending to my Father, to my God and their God."

There are two interesting aspects about this so far.

First, the undisturbed grave clothes, indicating Christ's body had passed through them when He rose.

Second, Jesus wouldn't let Mary touch Him because He had not yet presented Himself before the Father. Yet later that day, Jesus walked seven miles with two of His followers from Jerusalem to Emmaus. They didn't recognize Him either, until He broke bread at dinner. The Bible says, "Their eyes were opened, and they knew him; and he vanished out of their sight."[16] Jesus showed no concern now whether He was touched or not. Why?

The mystery deepens as we examine His appearance that night before the disciples.

Within an hour after Jesus vanished from the presence of His hosts in Emmaus, they left for Jerusalem. Mary had been quick to spread the news of Jesus' resurrection.

Fearing those who had crucified Jesus, and that they might be next, the disciples and other believers decided to meet behind locked doors that night to discuss the situation. The two from Emmaus were among them. As they told the others how Jesus had appeared to them on the road to Emmaus, and how they had not recognized Him until He broke bread with them that afternoon, Jesus suddenly materialized in the middle of the room.

The men were terrified. They thought Christ was a ghost.

"Why are you frightened?" He asked. "It's me! Jesus. Look at my hands! Look at my feet. *Touch me!* I'm not a ghost. Spirits don't have bodies, as you see me have!"[17]

As He spoke, Jesus held out His hands so everyone could see the holes from the nails. Then He showed them the nail holes in His feet.

The disciples were full of joy, but they still had their doubts.

"It's too good to be true . . . I just can't believe it. . . ."

"Do you have anything to eat?" Jesus asked. A playful smile crept across His face.

"Just some broiled fish. It's all we have left . . ." one offered.

Jesus took the fish and ate it. No one spoke; they just watched in awe.

"That was good," Jesus commented, wiping His mouth and hands on a napkin. "When I was with you before, don't you remember. . . ."

Jesus invited everyone to sit down as He began to open

their understanding of the Scriptures foretelling His death, burial and resurrection.

What a day!

There were no space ships in Jesus' time, yet He had traveled through perhaps millions of light years of space to Heaven and back in less than a day. He had walked seven miles from Jerusalem to Emmaus, eaten dinner, vanished into thin air and reappeared in a locked room in Jerusalem a short time later.

But Jesus made a statement that's more exciting than this. ". . . a spirit hath not flesh and bones, as ye see me have" (King James Version). Note: *flesh* and *bones*.

Jesus' new body didn't have blood. He had shed it on Calvary.

His body could travel through space as fast as a blink; could enter a locked room;[18] could eat; could appear unrecognizable to His closest friends, then change so it could be recognized;[19] could appear and disappear.

His mortal body had been transformed into immortality, into a body without limitations. The Bible says, "When He comes *we will be like him.*"

Today we are looking to that day of resurrection when we will be like Him. The creation of our immortal bodies will take place so quickly it will catch the world off guard. Our resurrected and glorified bodies will disappear from Earth, flash through space and leave this planet to its greatest period of tribulation.[20]

Others will be resurrected during the Great Tribulation predicted in Revelation. The two witnesses we've already discussed. But in the end of time there will be another general resurrection. This will be for the wicked. The ungodly of all ages will appear in their new bodies before God's throne of judgment, then go into eternal punishment.[21]

Today we have a choice of resurrections. Jesus said. "I am the way, the truth, and the life."[22]

The Bible says, "In Adam all die, even so in Christ shall all be made alive."[23]

To be recreated bodily in the resurrection of the godly, we must first have a resurrection in our spirit. To be in the physical kingdom of God, we must first be in His spiritual kingdom.

Resurrection is the key to both.

When physical death comes, all possibility of spiritual resurrection passes.[24] And without spiritual resurrection, there is only eternal death.

Resurrection in Church History

The resurrection is at the core of our confession of Christ for salvation. It has always been the center of Christian proclamation. It is the proof that God has really taken care of our sins and made it possible for us to come to Him. It made sure that the way to Heaven is open to us.

The Gospel, Paul says in I Corinthians 15, involves two things: Jesus' death and resurrection. These are always tied together. So we can be sure that any church or any period of history in which this truth is minimized or denied, has shifted the emphasis off the Gospel.

"He is risen!" was the spirit and substance of every greeting in the early church. And that's just the kind of emphasis that ought to exist in the church today.

Resurrections took place at various times during early church history to authenticate the Gospel. Irenaeus, Bishop of Lyons, wrote toward the end of the second century:

> Wherefore also those who are in truth His disciples, receiving grace from Him, do certainly and truly drive out demons . . . others again heal the sick by laying their hands upon them, and they are made whole. Yea, moreover, as I have said, *the dead even have been raised up, and remain among us for many years.*

And what more shall I say? It is not possible to name the number of gifts which the Church throughout the world has received from God in the name of Jesus Christ . . . and which she exerts day by day for the benefit of the Gentiles . . . by directing her prayers to the Lord . . . and calling upon the name of our Lord Jesus Christ she has been accustomed to work miracles for the advantage of mankind.[25]

Today, Jesus is again saying to the world, "I am the resurrection and the life. I am triumphant over the grave, and because I live, you live."

This is the message of the church that is Christ centered and Gospel rooted.

Nothing antagonizes the humanists more than the teaching of the resurrection of Christ. It cost Jesus His life when He declared that He would rise again. In the next chapter we will examine the historical fact of His resurrection and what it means to each of us.

10
Too Great for the Grave

The Prince of Life had been killed. Hell was dancing with glee, for it had conquered the Conqueror. Darkness had covered the Earth. It was not merely an eclipse, for it was full moon at noontime.

All Earth had given up hope. But Heaven knew better. Jesus was placed in the borrowed tomb, which He was soon to give back. Perhaps Gabriel said to the celestial choir, "Hold your breath . . . the world's greatest earthquake is about to take place."

Suddenly the earth shook. I call it the Easter earthquake. When Jesus was crucified, the veil of the temple was rent by an earthquake. Now the aftershock was even greater. Like a snap of a breaking seal, the stone began to roll back. Jesus stirred and stood. Like a butterfly slipping from its cocoon, so Christ shed His grave clothes.

The angelic choir, rising tier upon tier, burst into song.

"Good morning, Gabriel," Jesus said. "Tell the women who are coming to the tomb what has taken place. I'll see you a little later in my Father's house."

Many of the ordinances and teachings of the church reinforce the resurrection. Not only does history and the biblical record say Christ is alive, but present concepts of the church affirm it.

1. *Sunday is the day Christians worship.*[1]
   The Gallup Poll reports that 42 percent of the American people attend church each Sunday.

2. *Holy Communion.*
   "Upon the first day of the week, the disciples came together to break bread. . . ."[2] Each time Christians around the world celebrate Communion, they demonstrate their belief in a resurrected Lord.

3. *Prayer.*
   ". . . He everliveth to make intercession for them."[3] Christ's work on the Earth was redemption. In Heaven it is intercession. As our high priest, He delights in presenting our case to the heavenly Father.

4. *The Holy Spirit.*
   Each time a person receives the gift of the Holy Spirit, it is further proof that Jesus is alive. Our Lord's first priestly prayer when He ascended on high was, "I will pray the Father, and He shall give you another comforter, that He might abide with you forever."[4]

5. *Water baptism.*
   "Therefore we are buried with him by baptism into death; that like Christ was raised up from the dead . . . we shall be also in the likeness of his resurrection."[5] One

of the prime purposes of water baptism is to demonstrate the risen life to the world.

6. *Healing.*
"And with great power gave the apostles witness of the resurrection of the Lord Jesus. . . ."[6] Multiple miracles of healing brought followers to a faith in the risen Christ.

7. *The changed life.*
"And you hath he quickened, who were dead in trespasses and sin."[7] The changed believer who has encountered resurrection realities is a witness to a living Lord.

8. *Death.*
"And God hath both raised up the Lord, and will also raise up us by his own power."[8] St. Paul explains here that our victory over death is a direct analogy of Christ's resurrection.

Men Don't Die for a Lie

Skeptics scoffingly call the resurrection of Christ a mere myth born of tradition. arguing that the story was told and retold so many times it evolved into a celebration called Easter.

Eleven of the 12 disciples were martyred for preaching the resurrection. Millions through the ages have laid their lives down for this message. *Men don't die for a lie.*

It cost Jesus His life when He declared He would rise again. This was the basic accusation against Jesus when His enemies nailed Him to the cross.

What cost Paul's life? The same truth. When the apostle preached the resurrection on Mars Hill in Athens, the intellectuals mocked, but others wanted to hear more. Brought to trial for his faith, Paul proclaimed, "For the hope of the resurrection of the dead I am called into question."

About 32 percent of the 85 chapters in the New Testament that cover the last years of Christ's life concerns the resurrection. And more than 30 percent of the New Testament expounds the death and resurrection of Jesus, which proves the importance of this message to the early church.

The password of early Christians was, "He is not dead, but is alive!" The sign of the fish in the era of the catacombs was the symbol of resurrection. In the modern church this truth is rarely preached. Yet the resurrection message must be declared. Churches cannot remain neutral because it is the most significant issue of the Christian faith—even more important than the virgin birth.

I believe in the virgin birth, that the Holy Spirit came upon Mary, and Jesus was begotten by the Spirit. But God doesn't tell us in the Bible that it is necessary to believe in the virgin birth to be converted. It *does* say we must believe in the resurrection.[9]

Planned from all eternity, and predicted by the prophets. the resurrection is the heart of the Gospel, the whole alphabet of human hope and the best established fact in history.

Historical Fact

The resurrection is one of the best documented historical events. The biblical historian Edershim writes:

> If He was what the Gospels represent Him, He must have been born without sin, of a virgin, and He must have risen from the dead. If the story of His birth be true, we can believe that of the resurrection; if that of His resurrection be true, we can believe that of His birth.
>
> In the very nature of things, the latter was incapable of strict historical proof; and in the nature of things, His resurrection demanded and was capable of the fullest historical evidence. If such exists (and thank God it does) Jesus is the Christ in the full sense of the Gospels.

The Apostle Paul argues in I Corinthians 15 that if Jesus

did not rise from the grave, our faith, preaching, the church —everything—is in vain.

In this great resurrection chapter, Paul discusses the implications had Jesus not risen.

First, preaching is useless.

Second, we are then false witnesses.

Third, Christians are not only deceived, but deceivers.

Fourth, people are still in their sins.

Fifth, we are miserable.

The historical record of the Gospels is strengthened by the fact that in each account of the resurrection there is an absence of collaboration.

"These details witness to the freedom from disguise, to the artless frankness and candor of the writers," says biblical scholar Frank M. Boyd. "There is not the slightest evidence of deception."

Physiologically, the fact of Jesus' death is established in the flow of blood and water from His side when the soldier stabbed Him with a spear.

His death also is obvious from the record of His burial. Joseph of Arimathaea was permitted to take Christ's body for burial only after Pilate was convinced of His death.

Jesus was buried in a tomb that was sealed and guarded by Roman soldiers. Their lives depended on Christ remaining dead. But Jesus did rise, and He was seen by many after the event.

Some believe that the resurrection, which occurred 2,000 years ago, is remote to the twentieth century. How could a man's death and resurrection so long ago be relevant to me?

Jesus said, "Because I live, ye shall live also." In the first chapter I pointed out that most of the beyond and back cases mentioned in this book are firsthand reports. The resurrection of Christ is no exception. I have personally experienced the resurrection, and so have you if you've encountered Christ.

Paul, while not physically present at the scene of the resurrection, nevertheless declares that he was an eyewitness. The spiritual dimension of resurrection was so real to him that he identified with the 11 disciples as though he were there.

The Easter Parade

The first Easter Parade was not a flashy display of bonnets and bows, flowers and frills, and lace and linen. In stark contrast, the passing parade that first Easter was the more than 500 marchers who, according to Paul, passed the empty tomb and beheld the risen Lord.

What they saw not only changed them, but changed the world.

They Did Not See a Spirit

The appearance of Jesus behind closed doors was a spectacular event. His followers were certain He was merely a spirit. Jesus refuted this by exclaiming, "Handle me and see a spirit does not have flesh and bones as ye see me have."[10]

The composition of the resurrected body was supernatural; yet it was physical. He could walk through walls, but you could touch Him. He could eat, though He would never hunger. He vanished from the eyes of the Emmaus disciples, later to appear to the 11 disciples behind closed doors.

Belief in a bodily resurrection is an absolute. In the Greek New Testament the word *resurrection* always refers to the body—never the spirit or the soul. Careful analysis will reveal that all Christianity rests upon the fact of His bodily resurrection. This is why St. Paul records nothing about the resurrection of spirits or souls.

Although Christ's body is physical and has human appearance, its substance was changed, and it has spiritual characteristics. He has an "other-dimensional" body, one that can step from the spiritual world into the physical and back again. One that can step from Heaven to Earth and back without the limitations of time, distance or matter. One that can live in the spirit world where corruptible flesh and blood cannot exist.

Why is salvation predicated on confession and the resurrection? Because if Christ did not come back from the dead, He couldn't have been the prophesied deliverer, the great Messiah. The Old Testament prophecy was twofold relative to Christ.

First, He should suffer and die.

Second, He should reign.

Jesus was depicted both as the suffering servant and the coming King.[11] The Messianic prophecy that Jesus should be cut off from His people has already come to pass. So how could He be *cut off* and still reign, unless something intervenes? Answer: resurrection. He fulfilled David's prophetic words, "Thou wilt not leave my soul in the grave . . . thou will not permit thy holy one to suffer corruption."

What did the early church proclaim? The fact that God kept His Word. Messiah had died and was raised from the dead. Thus, because He is alive, He is indeed capable of judging the world. And if he is capable of judging the world, then quite obviously He is able to save it. So this became the theme of the church.

When the women at the tomb ran back to Jerusalem to tell the others, they were the first to begin spreading this good news of the Gospel. Since then, millions have believed and thousands have given their lives for this truth. Here is further proof that men will not die for a lie.

Death of Death

The resurrection, which pronounced death on death, has been described as the dressing room to immortality. The tomb was the place where Jesus laid aside the robes of death. And when we come to the end of our mortal existence, it can be the dressing room for the next life where we don the robes of eternity.

In the next chapter the efforts of modern science to prolong mortal life through machines and freezing will be examined. I'll discuss the controversial aspects of mercy killing: *who* has the right to decide when to pull the plug on the incurably ill? And *when* should life support be withdrawn?

11
Should Karen Be Allowed to Die?

One evening in April 1975, Karen Ann Quinlan went to sleep in her suburban New Jersey home. She never woke up.

Now more than two years later, as of this writing, the girl whose case has sparked heated debate over who has a right to "pull the plug," lies in a coma in Morris View Nursing Home in Morristown, New Jersey.

At age 23, suffering from a lethal combination of barbiturates and alcohol,[1] Karen Ann does not resemble her former self physically. Her body has gradually assumed a fetal position. Said to be in a chronic vegetative state with no hope of recovery, she has shriveled to a mere 60 pounds, half her formerly robust 120-pound self. Nothing remains of Karen Quinlan but a minimal degree of brain activity and biological life functions.[2]

For about a year the young woman lay in the Intensive Care Unit of Morristown County Hospital, two tubes in her neck—one feeding, the other helping her to breathe—while legal, medical and moral experts and family battled over her fate.

Karen's adoptive parents, who gave up all hope of her recovery, wanted to pull the plug on the respirator and let her die in dignity.

"Let us allow Karen to return to her natural state so that we can place her body in the tender, loving hands of the Lord," Joseph Quinlan begged the courts. "It is the Lord's will that she be allowed to die."

Their plea found sympathetic hearts around the world, even inspiring soul-searching poetry:

They hold me here

between two worlds

    The flesh leaves my hollow bones
    Even my breath does not belong to me

    I curl into a shape
    which I remember
    from another time

    But here
    my mother does not nourish me
    I am not sheltered
    in that secret room

    Mother?

    My eyes open and close
    but they look
    on another place
    where something
    waits for me

Why won't they

let me go?

—Caryl Porter[3]

Finally in a landmark decision on March 31, 1976 the New Jersey State Supreme Court gave Joseph Quinlan the power to remove his adopted daughter's life support system.[4]

Karen Quinlan—shown before her illness. She's still alive, in a coma, awaiting her fate. (United Press International photo)

Was it right for him to pull the plug?

Yes. If God wants a person to live, He can override the machines as He has done in the Quinlan case. Medical experts feared she would die immediately without life support. But as of this writing, Karen has been off the respirator for more than one year, and she is still alive!

Perhaps by the mercy of God her life will slip away. Or Karen may miraculously recover. However, she could live on in a deep coma for years. We don't know.

The Quinlan case has been a symbolic one. While the New Jersey Supreme Court has issued a precedent-setting opinion, this has not settled the debate. Karen Ann Quinlan will continue to captivate our attention as we deal with the difficult questions her lingering life has raised.

Right to Die

One of the leading questions being asked is, *does an individual have a right to die with a minimum of suffering?*

George Zygmaniak was only 26, and he was looking forward to a bright future until his motorcycle accident. But as he lay in a hospital bed in Neptune, N.J., paralyzed from the neck down, Zygmaniak felt he was a broken-down piece of machinery. So he begged his 23-year-old brother Lester to kill him. Lester did, using a sawed-off shotgun at close range, and was charged with first-degree murder.

Eugene Bauer, a 59-year-old cancer victim, was admitted to Nassau County Medical Center on Long Island one December. Within five days he was in a coma and was given only two days to live. Dr. Vincent A. Montemarano injected an overdose of potassium chloride into his patient's veins. Bauer died within five minutes. The physician listed the cause of death as cancer, but prosecutors said it was "mercy killing" and charged the doctor with murder.

The Whittemores were married during the war. He was a handsome, six-foot-two, medical corpsman on leave; she was a fun-loving Irish girl with deep blue eyes who captured his love. They had two children and never missed a Sunday in church with their family.

One June day Ray Whittemore blacked out at work and collapsed. When the diagnosis was made, Anne Whittemore could hardly believe it. Her husband was thought to have a tumor on his brain. When the surgeon operated, he found a malignant growth so massive that to remove it would have made Ray more vegetable than man. Ray, she was told, would not live much longer.

Eventually cobalt treatments burned off Ray's hair, and he would suffer agony with each treatment. Anne saw ahead to the hideous future facing her husband in the hospital cancer ward—bandages covering an emaciated body, tubes thrust into every opening, machines, needles . . . Ray had lost 80 pounds by the time he entered the hospital for his last stay, was blind and in severe pain.

Anne gave the doctors strict orders: "Do nothing except keep him comfortable and out of pain. No medical hocus-pocus." Keeping him alive didn't make sense anymore.

Early one morning a short time later, she received a call from the hospital, saying her husband had just passed away.

"Thank God, his suffering is over," Anne said quietly.

These cases underscore the growing emotional debate over euthanasia—mercy killing—and the so-called right-to-die with a minimum of pain for both the patient and the family.

Modern science has increased the doctor's dilemma in this area: how long to prolong life after all hope of recovery is gone and yet remain faithful to the Hippocratic oath, part of which reads: "I will neither give a deadly drug to anybody, if asked for, nor will I make a suggestion to this effect."

Artificial respirators, blood-matching and transfusion systems, heart and lung machines, and a myriad of other treatments can sustain a dying person's vital signs for weeks, even months, after his body has become incapable of living on its own.

Medicine's ability to extend life brings with it the ability to prolong dying. In using his skills, a doctor may thus be inflicting unnecessary suffering on the dying. And so the death with dignity debate gets some of its fuel.

The sanctity of life principle gets heavy billing in the debate. But does this require that we keep a patient's body viable when there is no sign of brain function? Are there certain cases of relentless and unrelievable suffering in incurable patients that is so intense as to make continued existence inhuman? Does the sanctity-of-life issue stand in the way of relieving suffering for which the only relief is death? Can human life be defined only in terms of biological viability? And must this viability be preserved at all costs?

Mercy killing is forbidden by law, moral tradition and medical practice. But it is receiving increasing support from physicians and the public.

A recent Gallup poll asked: "When a person has a disease that cannot be cured, do you think doctors should be allowed by law to end the patient's life by some painless means if the patient and his family request it?" Of the 53 percent who answered "Yes", younger and college-educated persons accounted for the larger percentage of the affirmative responses.

A recent issue of *Medical Opinion* reported that 79 per cent of the doctors responding to a poll on physicians' attitudes toward euthanasia expressed some belief in the patient's right to have a say about his own death. The majority of respondents said they would take no heroic measure to keep a terminally ill patient alive if there was no hope of survival. Only one in 10 said they would do everything possible, while 11 percent indicated they would take positive action to end the patient's suffering. Another 10 percent said they would choose a course of passive neglect, simply withholding supportive treatment.[6]

Dr. Bruce A. Becker, a Christian physician and regent at Melodyland School of Theology, shared his view with me recently:

Death is a transition for a Christian from life on Earth to life eternal with Jesus in glory. But for the non-Christian it is a gateway to an eternity in darkness, separation from God and eternal punishment.

Jesus said He came to give life, and I feel that depriving people of the opportunity for eternal salvation is equivalent to murder. I have seen, and there are instances reported by others, of patients having a true salvation experience while in coma with a terminal disease.

Only the Lord knows how many countless others have been saved while in a terminal, seemingly comatose or semi-comatose, state while on or off machinery and equipment designed to sustain life processes.

These machines only tend to sustain the physical signs and monitors of life processes such as blood pressure, heart rate, respiration, etc., but cannot monitor what is actually happening inside a person's heart, soul and spirit.

When the time arrives for the soul to slip into eternity, the Lord can and does stop the vital signs and causes that. soul to depart despite all the complicated medical machinery that we have in our armamentarium.

I believe man should do everything he can to extend life until there is no hope. But there's a point when one is clinically dead—when there is no longer brain activity—that the plug should be pulled. If God wants to intervene at that point, He can.

One of the critical spiritual issues of this question, *Does an individual have a right to die with a minimum of suffering?* involves one's views of what lies beyond. If this life is all there is, maybe we should cling tenaciously to any fragments of existence—even if it's mere vegetable. For the Christian, however, this life is only the prelude to a fuller expression of living beyond death.

When Dr. James Pippin, formerly pastor of one of the largest Disciples of Christ Churches in the Nation, spoke at Melodyland, he told about two sisters who kept their mother alive for 14 years:

> I want to talk with you for just a few minutes about eternal thinking. We don't live like we believe. We don't come to crises like we believe. We prate it, we talk it, we write it, we marquee it, we facade it, but we don't live it.

> Mamma got sick, and mamma got bedridden, and mamma had two daughters. This little ole mamma was a member of my church. A precious little old lady, she was 82 and active until she was struck down in bed. Know what happened?

> The daughters both had good jobs. So they said, "Well, one of has got to quit work." One of them said, "Well, you make more money than I do, so you work, and I'll quit. I'll stay home to take care of Mother."

> So she stayed home, and day in and day out she spoon-fed Mother. She rubbed Mother; she shampooed Mother's hair. She made up Mamma's face; she dressed and bathed Mother. She changed her sheets day in and day out and day in and day out, and would not let Mother die.

> And she'd come to church, and I'd say "How's Mother?" "Oh, she's just fine; she's going to live a long time."

> And they nursed Mother, nursed Mother, nursed Mother; 88, 89, 92, 94, 96. And they finally gave up and let her go to glory. What a pity. She was bedfast. Fourteen years earlier she could have gone to the Lord. But they wanted to keep her away from there.

> "The last place we want Mother to go is to the Lord, so I'm going to stay here and I'm going to nurse Mother. She's going to stay with us forever and ever and ever."

And if that sister didn't die two years later! Though she worked herself to death for Mother, she never got a decent night's sleep. Mother just slept like a baby, but she was up and down, up and down, up and down all night, looking to see if Mother was still breathing.

My Lord, have mercy. Let the people go to God! Let them go.

If I were dying, I'd say, "Don't put anything on me. Don't put an automatic breather on me. Let me go on to glory. Turn me loose, let me go." That's God talk.

Anybody who has been beyond doesn't want to come back. Yet when you come in to see the pastor after the doctor has told you, "You only have a few weeks to live," it's "Boo hoo, hoo. Pastor, somebody pray for me, I'm gonna die."

But, of course, you're going to die! What's so new about dying?

Let's get to thinking about living forever, but not in the flesh. Let's let the flesh go, and then we'll be free.

Christian psychiatrists tell us that the fear of death spawns every other fear, and if you could ever get over it and get into God's thinking about death, you'd be free. At death, your spirit is free.

When one's spirit goes to God, it certainly is not going to come back unless He wants to extend that person's life. I repeat, the Bible says there is an *appointed* time to die. Everything hinges on finding out *when* that appointment is.

Now what about the people who have come back from the dead? God's hour for them had not yet come. It takes the Holy Spirit to know the will of God relative to one's death date. I'll discuss this aspect later. But first we have another question to answer.

When a Person Dies

When is a person dead? The accepted theory for thousands of years was that death occurred when heart action and breathing stopped. Essentially, this was true because the brain died within minutes after the heart stopped. However, with machines it is possible to keep the brain "alive" almost indefinitely. When the brain ceases to function first, it is possible to maintain heart and lung activity artificially.

Legal definitions of death lag far behind medical advances. But in most cases, today's criterion is the absence of brain activity for 24 hours.[7]

It's my opinion that if there is no activity of the brain, physicians should remove life support machines. I would want them to keep the machines on me as long as my brain is alive because there would be time for a miracle.

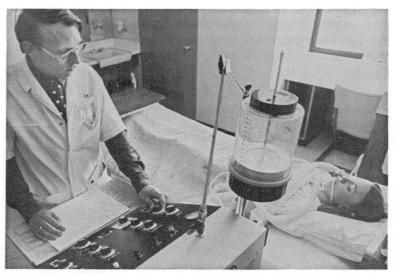

Sophisticated life support systems extend a patient's life. Should the plug be pulled?

When you work in a hospital and are around sick people, as I have been, you'll see cases where you thought the patients had technically expired; but because there was still brain activity, they were being kept alive by machines. There have been some amazing recoveries—even without prayer—in such situations.

A classic example of this was in the early days of our ministry at Melodyland. One of the deacon's wives was on life support machines for three days. A nurse, who worked at the Mayo Clinic and was assigned to attend the patient, told me when I called at the hospital, "She cannot recover. I have sat with many cases like hers, and they never come out of a coma when they're this far gone. I've been here for three days, and there's not even a quiver from her body."

After the hospital visit, I went to San Clemente to a home prayer group meeting. I opened the service by telling them, "I have just gone to see Lillian. There seems to be no chance of her making it. We should stand with the family in prayer."

Two of our church members who were at the meeting got up and left, saying they felt led to go pray for Lillian.

At the hospital, while they were praying, Lillian's eyes began to flutter. The nurse jumped up and became very excited because it was the first sign of life the deacon's wife had had in three days.

Within a few hours, Lillian awoke. She soon fully recovered and was released from the hospital.

Had the patient not stayed on the life support machines, she could have passed on before prayer prevailed.

When the Spirit Leaves the Body

Another important question is, *When does the spirit leave the body?* Does it leave when the heart and breathing stop? Or when the brain no longer functions?

The Apostle Paul said that to be absent from the body is to be present with the Lord. Some religions teach that for 72 hours after death one's spirit hovers over his body. We don't know this as a fact.

Many people have reported out-of-the-body experiences in which their spirits hovered close by while doctors frantically tried to revive them. Some of these were in operating rooms; others at the scene of an accident. In these cases, the heart and breathing had stopped, but brain activity was still present.

As long as the brain is alive, I believe the spirit is still there, and though it can be out of the body temporarily, it returns when breathing and heartbeat are restored.

When brain activity ceases, however, the spirit departs immediately.

Lethal Doses of Drugs to Relieve Pain

At the heart of the mercy killing debate is a concern over unnecessary suffering. The question often is asked, *Is it right for a physician to increase doses of painkillers to relieve suffering, knowing this will ultimately terminate the patient?*

Pope Pius XII, who was considered a stern guardian of traditional morality, declared that life does not need to be prolonged by extraordinary means. But like most other moralists, he said life must be maintained if it is possible by such ordinary means as feeding, usual drug treatment, care and shelter. However, for cases where the line is unclear between ordinary and extraordinary means, Roman Catholic theology says that if the physician's intention is to relieve pain, he may administer increasing doses of morphine, knowing full well that he will eventually reach a lethal dosage.[8]

I disagree with that. Jesus was offered hyssop, a drug used in Israel to relieve pain, when He was on the cross. Jesus refused it.

I would not want to be so drugged that I wouldn't know I was passing from this life to the next. I'm not fearful; Christians have nothing to fear.

How much better it would be to be coherent, to have your right mind, to be alert when passing through, to be able to communicate with your loved ones and know that your mind is not clouded with painkillers.

This doesn't mean one shouldn't be relieved of pain. I'm only talking about increasing doses to the point where the drugs become lethal.

Who Should Decide?

One of the most critical questions of the right-to-die debate is, *Who should decide when to pull the plug?*

At one time the church was the determining voice in matters of life and death. In the Old Testament the priest also was the physician. No leper could be called cleansed until pronounced so by the priest. Sanitation laws also were enforced by the priest. Then, as the history of medicine developed, the first hospitals were built by the church.

But because the church has relinquished its responsibility in this area, the medical profession now has the strong voice. It is a very sad, deplorable commentary on moral conditions

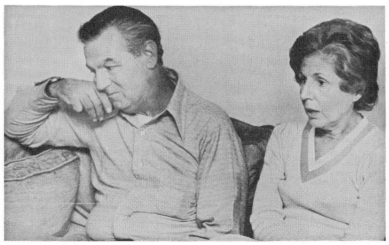

Mr. and Mrs. Joseph Quinlan ponder their decision: should they keep Karen Ann on the machine? (United Press International photo)

for the doctor to ignore the church and arbitrarily decide when and who should pull the plug.

Who, then, should decide?

Although the physician and other medical staff have the essential roles of determining what the patient's condition is, their expertise does not necessarily extend to all the dimensions of the patient's situation. In such life-and-death decisions, ideally the family, medical staff and clergy can reach a joint decision, with the family having the final say.

The pastor should be with the physician and family in prayerful consideration of the facts. Has everything been done to enable the patient to live? Is there any hope? Has brain activity ceased? What kind of existence will the individual have if kept alive artifically? Will he be more vegetable than human? Is he enduring intense pain? What needless suffering is the family going through?

The determining factor here is knowing in your heart that you have peace.* God will give you the assurance that you've done everything that one can do. The state has no right, neither do the physicians, to determine what should be done in critical cases like this. As a pastor and a family share together, they should tell the physician what they want done.

Sometimes the patient doesn't have a family. Possibly not even a minister. Then the physician has to play this role. And it's a heavy responsibility. Some doctors have become alcoholics or drug addicts because of the heavy moral responsibility they have to carry. What I'm saying is this: Let's relieve the physician. Let's assist him.

What concerns me is to see families burdened with exorbitant doctor bills which they can't get out from under, and to see them torn apart emotionally.

Some have sat by the bedside of the sick one for months, to the point of collapse, knowing there is no possibility of that person coming back without divine intervention.

God doesn't need a hook-up. He can do without it. And

I do not accept the charge that pulling the plug is murder if brain activity has ceased.

How beautiful it would be for a praying physician to enter into the conclusion with the family and minister. More and more we're seeing physicians and the Christian minister coming together and resolving these problems. The medical profession and the Christian ministry need to come closer if one is to cope with the fantastic new problems we're all facing in this scientific age.

Living Wills

The wishes of the patient should certainly be taken into consideration. Many thousands of healthy Americans have signed "living wills" like the one prepared by the Euthanasia Educational Council. It states:

> If the situation should arise in which there is no reasonable expectation of my recovery from physical or mental disability, I request that I be allowed to die and not be kept alive by artificial means or "heroic measures."

> I do not fear death itself as much as the indignities of deterioration, dependence and hopeless pain. I therefore ask that medication be mercifully administered to me to alleviate suffering even though this may hasten the moment of death.[10]

In effect what the document says is, "If I'm terminally ill, pull the plug." While this is not legally binding, it is of considerable help to those who must make the decision. Privately, doctors say they often respect the will.[11]

If the individual is of a rational mind when he makes the "living will," we should not override his personal rights—providing there is a cessation of brain activity.

Freezing the Body

The bodies of 15 people have been frozen in liquid nitrogen in the hope that someday they can be revived when medical science has found a cure for the disease that killed them. *Is putting death on ice a gateway to immortality?*

The idea of freezing the body—called cryonics—was first proposed 13 years ago by Robert Ettinger in his book, *The Prospect of Immortality.* His book sparked the formation of cryonics societies across the nation.

A physics professor from suburban Oak Park, Illinois, Ettinger considers cryonics an extension of what medical technology has been trying to do for years—extend life and conquer so-called natural death.

Although 15 persons are awaiting bright tomorrows, about 30 had their bodies frozen at one point. Interest in the idea has been thawing, however, and coupled with the high costs, relatives of about half have opted for traditional burials.

The freezing process begins immediately after death. The body is gradually cooled in ice and the blood replaced with a special chemical solution, a process similar to embalming. Ice is used to cool the body further and a second solution replaces the first. The body temperature is then lowered slowly to minus 320 degrees Fahrenheit with liquid nitrogen vapor, and the body stored in a thermos-shaped capsule, bathed in liquid nitrogen.

While Ettinger hopes to cut the costs drastically, the average bill for the death freeze is about $50,000.

Before freezing can begin, the body must be legally dead, however, requiring an assortment of documents, which must be signed well in advance.

The 15 who are now in deep freeze believed that someday one of two things would happen:

1. A way would be found to bring the body back to normal and cure the disease that killed it.

2. Technology will be developed to grow an identical body

from the genetic information in a cell nucleus of the original body.

The latter hope assumes that by the time such a technique is developed, the physical basis for personality and memory will be well enough understood so the person's identity can be implanted in the brain of the new body.[12]

While we cannot say cryonics is unscriptural, putting death on ice in hopes of immortality poses some unique problems.

First, time is extremely critical. Once a body is pronounced clinically dead, it must be frozen within four minutes if brain damage is to be avoided.

Second, the theory that cellular destruction will not take place in the central nervous system of the brain is based on the premise that the heart can be reactivated by electrical charge. This has never been proved, except on the operating table while the body is still warm. Cryologists admit ice crystal formation in body cells is a real threat, doing irreversible damage to the delicate microscopic cells of which we are made. And how is one to know whether the freezing capsule will be maintained properly?

Third, none of the frozen bodies has been revived, so we don't know that one can be brought back to life.

Fourth, assuming it is possible to come back in a hundred or so years after a cure has been found, what kind of world would it be? All of your friends would be gone. You wouldn't know how to relate psychologically to your new society.

More important, however, you must assume when energizing the body that the spirit that activates it is still there. Man is body, soul and spirit, and whether the person is a Christian or a non-Christian, his spirit goes to its destined resting place, which is determined before he dies. The Bible says, "It is appointed unto man *once* to die, and after that the judgment." God is not going to call your spirit back from either Heaven or Hell because someone invented a way to preserve your body and bring it back to life.

Hollywood has created many grotesque creatures for the movies. But what is even more frightening than these suspense thrillers is that in experimenting with your body, scientists of the future could turn you into their version of a soulless Frankenstein monster, a hopeless captive of the "undead."

Exceptional Children

No matter how advanced medical science has become, thousands of babies are born each year physically and mentally deformed. In many cases, the damage is so severe the children have no chance of leading meaningful lives.

If an infant is born with a severe mental or physical defect, should it be allowed to live?

Many times I have visited hospitals where I've seen waterhead babies and other abnormalities that were unbelievable. I still believe that it's better to keep those children alive, and I'll tell you why: they are human beings; they are God's creation. They were born into the world with problems, but those precious babies can give incredible love.

Opera star Jerome Hines has a handicapped boy. Once, when visiting Melodyland, he explained the dilemma of having an exceptional child. Most parents would ask, "God, why did this happen to us?" He said, "God must have trusted us very much, to place a handicapped child in our care." He went on to express the immeasurable amount of love they had received from this child and the blessing he is in their home.

But what about bodies who cannot give love, who are hopeless basketcases? They are merely biological blobs with no capacity for a meaningful life or outward expression of love. And what about Heaven? If it is such a wonderful place, why keep them here in distortion and misery? Why should parents and society be burdened by these helpless creatures? Should mercy killings be allowed in these cases?

No. And for some very good reasons.

First, if rightly interpreted, every child can be loved, even if he is unable to give love.

Second, we have observed phenomenal response in some of the cases here at Melodyland in our "Overcomers" program for exceptional people. One of the prime successes of the ministry is teaching a handicapped person to help a more handicapped individual. The sheer gratification and pleasure of this accomplishment often brings progress in their cases.

The third reason is no one has the right to take another person's life. It's God who gives life. It's God who takes life. And that mass of humanity is valuable to Him.

Often parents blame themselves for some sin, believing this is why they have a handicapped child. But God doesn't create anything imperfect to punish wrongdoing. Imperfection is a *scar* of sin. Often the parents are godly, yet this misfortune occurs.

Organ Transplant

Are transplants of vital organs an alternative to natural death? If we're concerned with extraordinary measures for extending life, what about the one who is suffering from a heart attack? He is being kept alive by a machine, but the heart of a healthy person killed in a car accident is available. A decision must be made immediately. Should the transplant be made to save the man's life?

We are living in a scientific age when man has developed a lot of new technologies. We need to take advantage of every one of them. Here again, the family must decide. The church has nothing to say on some of these issues; it can only speak where the Bible speaks, and the Bible doesn't say anything about organ transplants. I can see nothing wrong in this, however, as an alternative to death.

Perhaps some would judge the church for speaking out on these vital issues. However, no longer are we dealing with merely philosophical problems, but rather with practical questions that must be faced daily.

A few years ago the church did not consider these areas because sophisticated knowledge and sciences had not advanced that far.

When a life is spared, either through miraculous healing or through a back-from-the-beyond experience, it raises many questions. Why did my best friend die and I live? What happens to a person five minutes after he dies? When death comes, do I cease to exist? Will I lapse into unconsciousness and sleep until a time in the future when I'll be resurrected? Or do I go immediately into a better life in the beyond? How can I conquer my fear of dying?

The next chapter will answer these questions and help you understand the meaning of death.

12
Death—Fear of All Fears

"Do something! The man is dying!"

On the verge of panic, I frantically pushed the button for Mrs. Wilton, the only other nurse on duty at the hospital. But she was in the bathroom of the women's ward and did not see the flashing light to respond . . .

Although it happened during my first year of college when I was only 17, the initial memory of that death scene has never been fully erased. Instantly, myriads of thoughts flooded my mind, some a replay of comments I heard about death during my childhood. But I instinctively responded to my first impulse, "Get help fast! The man is dying!" The next mental flash was even more crucial—*What happens five minutes after death?*

I was working in the emergency ward on the night shift at

a small, 50-bed hospital that served the community where I attended school. The dedicated surgeon who owned the hospital had only two interests left in life after his wife had died. His only son, an Army doctor, was stationed overseas. The first interest was his hospital; the second, his church, and precisely in that order.

When applying for the job at the hospital, I offered to do anything, even the unpleasant duties, because I was helping humanity.

The doctor said to me, "Son, since you would like to be a medical missionary, I'm going to let you learn by doing. Your assignment will be to assist with the night calls and emergency."

At two in the morning a light flashed from the men's ward. Only two men were in the room, and only eight ladies were hospitalized in the women's ward of the tiny hospital. As I walked into the ward, the frantic shouts of one of the men sprung me into action. "My God, do something! The man is dying!"

"Mrs. Wilton, get in here. Help me!" I wanted to cry out. But I was on my own for what seemed eternity until she finally spotted the flashing light and came bounding into the ward. Meanwhile, I had heard the faint whisper of the man's last words, "Do something . . . I'm dying . . ." Then, he gasped big, and it was over.

Soon the doctor arrived and pronounced him dead. The man, in his forties, was the victim of quick pneumonia, which had set in without warning. Putting away his stethoscope, the doctor said, "Well, this man has no close relatives . . . maybe it's best he went this way."

However, that night I couldn't shake the memory. It was my first time to see someone die. I kept asking myself, "Did he go into an unknown land called the 'beyond'?"

What happens when I die? Do I cease to exist? Is this life all there is? Do I lapse into unconsciousness and sleep until sometime in the future when I'll come back to life? Or does death release the real me to a better life in the beyond? After death, will I know what's going on in this world?

These and many related questions have been asked by all of us at one time or another. This chapter will give the answers. But first, let's examine some of the truths about death and how Christians see it.

Face the Facts

First, death is indiscriminate. It claims the rich, the poor, the aged, and the young. The good and the bad—all must face death. Ecclesiastes says:

> There is an appointed time for everything. And there is a time for every event under heaven—a time to give birth, and a time to die. . . .

Because of phenomenal advances in medical technology, nearly two-thirds of the deaths in industrialized countries are now associated with the infirmities of old age.[1] In its endless struggle against this age-old enemy, medicine has succeeded in fending off death for a longer period.

For example, physicians can substitute for all the life-supporting functions of the nervous system. Tubes into a vein or the gastrointestinal tract can take the place of eating. A similar tube into the bladder takes the place of normal elimination, and an artificial respirator can take over breathing. A variety of electronic devices can even keep the heart beating for months after it has stopped.[2]

But it is clear that we must all die sometime. And a healthy attitude concerning this ultimate event is absolutely essential if the last years of one's life are not to be spent in fearful misery.

Dr. Elizabeth Kübler-Ross has identified five typical stages of reaction people go through when they realize they're dying:

denial, anger, bargaining, depression and *acceptance.* It is in these areas that a healthy Christian view of death can help that individual and his family better cope with this ultimate experience.[3]

Let me improvise on these concepts:

NOT ME   It is very difficult for a patient to acknowledge he is terminal. "It can happen to others, yes, but not to me!" It is well to be honest with the sick if they are able to accept the facts presented in a loving way.

WHY ME   It's common for a terminal case to become bitter and reactionary.

YES ME   The patient now knows he is going, but bargains for more time. He may even become very depressed and not wish to see anyone. This may be a sign that he has finished his unfinished business.

NOW ME   It is not resignation, but victory. A time void of feeling, it is a glorious entering in for the Christian.

Believing in life hereafter is of infinite help to the terminally ill person and his family; for in death they have hope and can face the fact with less pain.

Second, death may be delayed, but not denied. At times it appears that medical science may wipe out all disease. Many specialists claim that by the year 2000 man will live to be a hundred. But if one lives to be 1,000, death would only have been delayed.

Robert Morison reports some interesting statistics on life spans and how medical science has been able to increase them:

At the beginning of the century, death came to about 15

percent of all newborn babies in their first year and to perhaps another 15 percent before adolescence. Nowadays fewer than 2 percent die in their first year and the great majority will live to be over 70.[4]

He notes, however, that life expectancy at 70 has changed very little.

The difference is that in earlier times, relatively few people managed to reach 70, whereas, under present conditions, nearly two-thirds of the population reach that age.[5]

Most people, he adds, no longer die of quick or acute illness, but of the chronic deteriorations of old age.

We have observed in the last chapter how modern technology is attempting to freeze bodies in an ultimate effort to preserve life. Even if cryonics succeeds, even if it were possible for the spirit to be called back from its resting place in the beyond, death would be inevitable; the individual would only die again.

Third, death determines destiny. No place in the Bible suggests that one will have a second chance beyond the grave. One's place in destiny is settled in this life. After death, there is no possibility of change of address.

In a realistic sense, we *now* have either eternal death or eternal life. At this very moment you are living in eternity. This is why it is often said that death is merely the comma in the sentence of life—there is more to follow.

What a fantastic thought: we are creatures of eternity. Ecclesiastes says, "He hath set eternity in their hearts." John Newton, the theologian said, "We are still in the land of the dying, and we will soon be in the land of the living."

Fourth, death is sometimes predetermined. Twice in the Bible God told men exactly how long they would live. I am certain, however, that some die before their time.[6]

There are two ways to kill yourself. First, at this moment

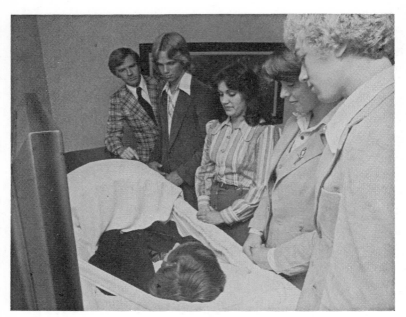

Students learn how to face the father of fears—death.

you are in the process of either living or dying. Your attitude toward the benefits of the crucifixion and resurrection determines the direction. Paul warns in Corinthians that some are sick and die because they do not discern the Lord's body. Second, suicide may be committed by abusing one's physical body. Improper rest, deficient diet or gluttony and assuming unrealistic worries take a tremendous toll. This is not God's plan. The Bible declares your body is the temple of the Holy Spirit.

   Fifth, death is a door. Just as one has a birth date, he has a death date. The three most important documents of a person's life are his birth, marriage and death certificates. And as birth is a door to natural life, so death is the entrance to eternal life.

   Strange, isn't it, that we are jubilant at the time of birth, yet often are near hysteria when someone dies. A newborn baby comes into a world that is filled with tribulation and trials; a believer in Christ who goes into the beyond enters an eternity of rest and bliss. Should one weep at death or when a baby is born?

Children, who have not picked up adult fears, usually have the purest insight into life beyond.

When my daughter Debi was 16, her high school had an occupational day for seniors. One-day honorary jobs, such as bank president, fire inspector and newspaper editor, were offered by business firms and organizations. Debi applied for the most bypassed job on the list—mortuary director. Here's her letter of application:

Dear Mr. Brown:

I am applying for the job you have listed at our school, mortuary director for a day.

My interest in this job is for a good reason. Most people are totally uneducated about death. Our system only teaches us how to live.

Just mention the word "death," and people react with shock. Most friends my age have never been to a cemetery except on Halloween night.

Being a P.K. (preacher's kid), I have seen the need for people to accept the fact that death comes to all . . . and to all ages. I have heard my dad preach several funeral sermons, and he always nails down one point . . . death is certain for everybody, but it is not the end. A better life is ahead.
Thank you very much.

Sincerely,

(Signed) Debi Wilkerson

Debi, of course, got the job, and on arriving at the mortuary was joined by two more students from other schools. They were first taken to view the body of a little boy before it was to be moved to a Catholic church for Mass.

Debi tells it like this:

> Dad, as the three of us stepped to the casket, one of the girls started sobbing softly. The little boy, about seven years old, was adorable.
>
> But the casket veil and the makeup did not hide the tiny sores that showed faintly on his lips. He must have had a long illness or a raging fever.
>
> Daddy, as I stood there I almost burst into tears, but suddenly I thought of the boy as just a small, well-dressed mannequin. The real little boy was already running and playing on golden streets.

On another occasion, Debi exhibited her lack of fear concerning death. When she was small, she loved to brush and comb her Grandpa Wilkerson's hair. He enjoyed these times when his granddaughter would climb up on a chair and stroke his shiny, black hair. Often I heard her whisper in his ear, "Grandpa, you are so handsome. You look just like a teenager." Any grandpa would enjoy that kind of attention!

My father went to be with the Lord when Debi was 13 years old. When the family went to the mortuary to view his body, Debi slipped over by her grandmother. After a while, she said, with a studied look, "Grandma, Grandpa's hair doesn't look like him at all."

"Honey," Grandma said, "the mortician asked for a picture of Grandpa to see his hairstyle. The only one I could find in a hurry was an old picture."

"Grandma, I remember how Grandpa wore his hair. I used to brush his hair when I was little and I think I can comb it like he wore it."

"Please do, Debi. Grandpa would be so proud of you."

After everyone left except our immediate family, Debi gently brushed Dad's hair into place perfectly. She was reacting the same as in childhood when standing on a chair to reach her grandpa's hair. It seemed natural for her to want Grandpa to look his best—even in death.

The Bible talks about three kinds of death. The first is temporal, where the spirit is separated from the body. The second is spiritual, which is separation of the spirit from God. It is the second death that is cast into the lake of fire, which then becomes eternal—the third.

> And death and Hades were thrown into the lake of fire. This is the second death, the lake of fire.[7]

For the Christian, death is only temporal; it is but the door to spiritual and eternal life. Death's destiny has no claim, for the bodies of those who have died in Christ will be raised to immortality.

Meanwhile, however, death does have a sting. And only those who have lost a loved one, spending many lonely hours, can fully understand the pain of loss.

When a Christian dies, those who are left behind do not sorrow as others who have no hope. This is not to say it is wrong to cry. It is important not to repress these feelings. Jesus wept over Lazarus' death. Mary and Martha didn't believe Jesus understood until they saw Him weep. Feelings of sorrow are natural and should be released, but excessive emotions are sometimes indications of guilt.

A lady in our church was stunned by the sudden death of her 40-year-old husband. He was obviously the picture of good health. He had completed a term as mayor of one of our neighboring communities.

In giving her words of comfort, it seemed the Holy Spirit prompted me to say, "You must accept comfort just as you accept salvation. Comfort is yours, but you must receive it."

Some feel that prolonged grief indicates continued devotion and commitment to the deceased. If the loved one is with the Lord, all your weeping isn't going to bring him back. Extended grief is merely martyrdom that will not change the situation. The loved one is not helped by your broken heart, for he is surrounded by celestial joy.

When a Christian goes beyond, why not remember him as he was in life? Keep the burial simple. There is no need for an open casket unless requested. Most people prefer to be remembered in their natural state. Even more so is this true in the case of prolonged illness or disfiguring accidents.

Often people unconsciously believe that by viewing the dead they can establish identity with the deceased. It was a common occurrence in past years for people to have portraits made of loved ones in caskets. Often pictures were inlaid in grave markers. Elaborate gold funerary masks were a ritual for ancient kings. In Egyptian culture, a king started building his pyramid for death when he took the throne. Most of the wealth that could be amassed during his lifetime was designated for his tomb.

I am firmly convinced that the casket should be at the service and not buried beforehand. So often the passing of a loved one seems like an illusion, and the family must realize that he is departed.

The funeral for a Christian is his coronation. Let the service be warm, with the congregation singing a favorite hymn. And though there is triumph, there is a place for godly sorrow and profound reverence in the service.

For those of us who contend for Christian healing, there comes a major paradox. How can one reconcile the teaching, "It is God's will to heal all that come to Him," with the death of a loved one? Often this question shatters a bereaved person's faith.

As my father lay dying, I continued praying for his recovery. The last night, while standing by his bed, the Lord gave me a verse in the Bible that has since helped many others. In Hebrews 11, Paul wrote, "These all died in faith, not having received the promise." Then God spoke to me and said, "Would you rather your father die in unbelief than in faith?" We must be quick to admit that no one understands fully God's will concerning the time of death. The question is not *whether* a person is healed, only the *timing*. My father has gone beyond, but he wasn't denied healing. It will come in the resurrection of the righteous.

What comfort it is to realize that in comparison to eternity, we are separated from our loved ones only a brief time.

Is Cremation Christian?

The question has been asked many times, *Is cremation Christian?*

The Jews cremated their first king, Saul, and buried him probably in a jar. They wouldn't have done this had it been forbidden by the law of God. Early Christians, however, did not practice cremation. They simply buried the body. The usual position of the Christian church is burial, not cremation.

Dr. Walter Martin comments:

> Apparently from all the literature we are able to see, nobody really got too upset about it.

> I would inject a rational argument here, a logical argument based on something Donald Barnhouse taught me. He said, "Dust thou art, to dust thou shalt return. The divine curse reduces men to dust. All the devil is going to get is a mouthful of dust in the end. Now what difference does it make if the body goes to dust in 100 years or eight hours?"

> The disintegrating body we place in the grave certainly isn't the basis for the body that comes forth in the resurrection. Our present bodies are sown in corruption, to be raised in incorruption; sown in mortality, to be raised in immortality.

It's pretty obvious, then, that the bodies we have now will have no bearing whatsoever on those that will be.

The New Testament does not make a statement on cremation.

What Is Man?

God's Word says man is composed of three parts—body, soul (mind) and spirit. Only the body is mortal; the soul and spirit are eternal.

Man became a living being when God breathed into Adam, forming him out of the earth like a sculptor would form a statue. After God created Adam in His image, He made him a reflection of Himself—an eternal spiritual entity.[8]

It is this part of man that lives forever; it is the body that returns to dust.[9] Death in the spiritual sense is not the end of one's existence. It is separation. For the Christian, then, dying is simply separating from the body to live in the beyond. For the non-Christian, it is more. Death also means being separated from God.

Man is a unity of body, mind and spirit. Some have tried to weigh the human body after death to see if anything is missing, if there is any weight loss when the spirit departs.[10] David Winter says:

> One might as well try to weigh a sense of humor, or a feeling of depression, or a passionate love. It speaks volumes for the crude materialism of man that he should for one moment suppose that *reality* is the same thing as *weight* or *volume.*
>
> Man is *personal.* He has personality and he is a person in bodily form.
>
> I conclude that, when I die, my body disintegrates, but my personality—the real me—lives on; this personality is going to express itself in the "life beyond" in a totally different kind of body.[11]

What Happens at Death

Now, let's examine what happens when one dies.

The dust will return to the earth as it was, and the spirit will return to God who gave it.[12]

When death comes, the spirit leaves the body.

Wernher von Braun, the famous scientist, says, "I believe in an immortal soul. Science has proved that nothing disintegrates into nothingness. Life and soul, therefore, cannot disintegrate into nothingness, and so are immortal."

The scientific world has no instrument to measure spirit, so for insight into this entity, we must turn to God's Word. But we can also glean knowledge from the testimonies of those who have died, gone to the world beyond and returned to describe it.

Not only does the Bible teach it, but we can see from the experiences of Marvin Ford, Dr. Harrison and others mentioned that the spirit goes into the beyond when it leaves the body. It does not cease to exist, nor does it go to sleep, as some teach.

According to the Bible, the spirit of a Christian goes immediately into the presence of the Lord. Yet, there may be mitigating circumstances in which the body may die and it may not be God's time. Or God may want to communicate something to the individual. That being the case, and there is evidence for this, a person can be given messages in a disembodied state, then return.

However, similar phenomena occur in occult experiences. Dr. Walter Martin, author of *Kingdom of the Cults* and a professor at Melodyland, says, "We have known it from the world of the occult for many, many centuries. Right now people are getting information from books and periodicals on documented cases of individuals who have left their bodies at the time of an operation or other kind of trauma, and come into contact with a being of beautiful light, warmth, peace and love. The entity, in some instances, confirmed to them the love of God or how they were "accepted."

In other cases an evil being would say it isn't really essential to live an absolutely perfect life and that they were acceptable to the deity just as they were.

"What you get is a contradiction of Christian theology through occultic experience," Dr. Martin continues. "Then the person returns to the body and tells about his experience. The *National Enquirer* has reported a hundred such accounts and I've read about these cases in books for 25 years.

"If this glorious being you encounter says something contrary to the Gospel of Jesus Christ, then you need to remember there is a glorious being known as the light bearer—Lucifer—who appears as a messenger of light. He really is a force of darkness arrayed as light, garbed for deception, who leads a person into thinking that he will have a peaceful life in the beyond simply by being good to his fellow man and living the right kind of life."

Whether an out-of-the-body experience is Christian or occult can be tested by the Word of God and the individual's life afterward.

The devil, who can appear as a being of light, cannot harmonize with biblical truth and is so exposed. The individual who has an encounter with Christ can never be the same, as we have observed in previous chapters.

Soul Sleep

Some teach that the soul merely goes to sleep when the body dies, awaiting the day of resurrection. We have already seen that this is a false teaching. But how did the doctrine evolve? What about Bible verses which use the word *sleep* as a synonym for death? And how does this teaching conflict with the experiences of those who have been beyond and back?

Dr. Martin says:

> The doctrine of soul sleep developed because people could not face up to the concept of Hell. Their idea of justice didn't permit Hell. So soul sleep was introduced as an intermediate stage and and was based upon certain misinterpreted passages in the Bible.
>
> I have covered most of them in *Kingdom of the Cults* in the chapters on Jehovah's Witnesses and Seventh-day Adventism. I've given all the references, the meanings of the words, the exegesis of the passage and where they are wrong.
>
> But to deal with soul sleep you have to recognize that the Bible does use the word sleep as a synonym for death. Jesus referred to Lazarus, for example, as being asleep when He was on His way to Bethany.
>
> "Oh, that is wonderful, Lord," the disciples chorused. "He must be feeling better and is taking a nap."
>
> Jesus said, "Lazarus is dead." So Jesus made the equation: sleeping is death. "I am going to awaken him out of his sleep."
>
> Those who believe in soul sleep and promote it vigorously cite Daniel, "They that sleep in the dust of the earth shall arise." And the Psalms, "The living know that they shall die . . . the dead know not anything."
>
> Next, they take Ecclesiastes, "What is the difference between the spirit of man and the spirit of the beast . . . one goes down . . . the other goes up . . . there is no permanence . . . they both die. Also, I Thessalonians 4, "Those who sleep in Jesus, God will bring with him."
>
> So you see, the word sleep in the Bible does have certain connotations where death is concerned. But when one sees that in physical death the body simply takes on the appearance of sleep, the equation of the two emerges as an

Hebraic figure of speech. The secret to understanding all this, however, is to take each passage in its context and find out what the word means there—then go to all the rest of the passages and put them together.

That is what I had to do, and their arguments disintegrated. On the surface, their arguments look very formidable when they are thrown at you just one verse at a time. But let me give one answer that is devastating to the soul sleep people.

The greatest of all the apostles is acknowledged to be Paul. He wrote the great body of the New Testament. What was his attitude toward death on a personal level?

In Philippians he says, "For me to live in Christ . . . to die is profit." There is no profit in soul sleep because you simply lapse into unconsciousness until the resurrection.

So for him to say "For me to live is Christ and to die is profit" doesn't make sense! It is better to go on living than to go into unconsciousness. This is one point.

Point two, Paul says, "I am torn between two things. I really don't know what decision to make. I long to depart and be with Christ, which is far better." Now, the word *depart* is an old military term, which means to weigh anchor or break up the camp and move out.

Paul was a grammarian, and it is marvelous that the Holy Spirit led him to use this word to describe what happens to you when you die.

This is the way it reads in the Greek text: "What I am longing for is the departing and the being with Christ." This is all participial. You can't separate them and say, *depart, go to sleep,* and then be with Christ because the participial construction will not permit it. Paul says, "I'm longing to break up camp—death—and be with Christ, yet to remain with you is more needful."

You see the contrast. If he were talking about the second coming of Christ, he wouldn't say there were two choices—to go or to stay with them—because they would be going with him.

Paul is trying to tell them, "Look, in the interim between now and Jesus' return, this departing is being with Christ." He reinforces this theology throughout his writings. II Corinthians 5 says, "To be absent from the body is to be at home with the Lord."

Paul is a man getting ready to die, and he asks, what am I looking for? Sleep? No. I'm looking forward to being with Christ, which is far better.[13]

Baptizing for the Dead

Another falacy is baptism for the dead, which in effect says, "Wait until you die and let someone else atone for your sin through baptism."

This teaching is based of I Corinthians 15:29:

What will those do who are baptized for the dead? If the dead are not raised at all, why are they baptized for them?

For several reasons, baptisms for the dead are not biblical.

First, nowhere in the Bible is one told to be baptized for the dead. Not only is there no command, nor biblical example, but the Corinthian church had many heresies, which Paul had to correct.

Second, in no case in the Bible does baptism save a living man. Anybody who repents of his sins and receives Jesus, follows Christ in water baptism. But one cannot be converted merely through baptism.

Third, every fact or doctrine should be proved by two or more Bible verses. I Corinthians 15:29 is the only verse in the Bible where the subject is mentioned, so if this were biblical practice of the early church, other verses would have to support it.[14]

Fourth, salvation is a personal matter. No one can die for our sins except Jesus Christ. Salvation cannot be obtained by proxy, so no one can be baptized for us.

Fifth, Paul here is showing the inconsistency of the false teachers in Corinth who questioned the doctrine of the resurrection, yet accepted the practice of baptism for the dead. Corinth was one of the most vile cities of its time. Many false teachings from heathen worship crept into the church, including baptism for the dead.

The force of I Corinthians 15:29 is in the argument, "If you don't believe in the resurrection, why bother baptizing the dead?" Just because Paul made a passing reference to the error doesn't mean he sanctioned it.

Prayers for the Dead

A companion error is praying for the dead.

The idea of praying for the dead was accepted by many during World War I. One book written during that period used the argument, "If praying be the mightiest weapon placed in our hands, we dare not restrict its power merely for the aid of the living. For the dead also are still on the same great stream of life as we are."

This practice goes back much farther in history than this, of course. The Jews prayed for the dead, not because of biblical revelation, but out of tradition encouraged by Judas Maccabeus in the second century B.C. An apocryphal book not included in the biblical canon, Second Maccabees, says:

It is a healthy and wholesome thing to pray for the dead.

The Roman Catholic Church canonized this during the council of Trent when it readmitted the books rejected by the Jews. It then was passed down in Orthodox and Catholic traditions. But the accepted canon of the Bible excludes the apocryphal books and thus this teaching.

Dr. Martin has these thoughts on the subject:

> The Bible says nothing about praying for the dead for two reasons.
>
> One, those who have departed believing in Christ are absent from their bodies and present with the Lord.
>
> Two, those who have departed and did not believe in Christ are "under punishment until the day of judgment" (II Pet. 2:9 Greek). Luke 16 says the departed unbelievers are where? In Hell.

An alleged example in the New Testament of prayer for the dead is II Timothy 1:16, 18:

> The Lord grant mercy to the house of Onesiphorus for he often refreshed me, and was not ashamed of my chains . . . the Lord grant to him to find mercy from the Lord on that day.

However, there never was a question in St. Paul's mind that this departed Christian was with the Lord—if he was departed at all. There is not the slightest proof that Onesiphorus was dead. Numerous church authorities say the practice of prayer for the dead finds no other sanction in the New Testament.

Conquering the Fear of Death

For many, fear is a paralyzing numbness that defies reason and cripples lives. Thoreau wrote, "Nothing is so much to be feared as fear." The philosopher was speaking of a type of fear that becomes debilitating—the kind that overtakes an individual's rational mind, that paralyzes the limbs, makes the heart race and the stomach twitch with nausea.

Such phobias account for the lady who has been afraid to leave her home for 28 years, and the woman who is so terrified of toy balloons she won't take her children to the park because somewhere a balloon may be spotted. Then there is the electrician who walks up 17 floors a day because he is afraid of the elevator.

Phobias like these cause jobs to be lost, marriages to dissolve and friends to be alienated. And thanataphobia, as the fear of death is called, itself can become a torturing experience.

Dr. Rex Rook, medical director of the Melodyland Psychiatric and Counseling Clinic, says the father of all fears is the fear of death, which began to plague man in the Garden of Eden.

The first person to investigate the relationship between Christian faith and attitudes towards death was Dr. Len Cerny, a psychological researcher who has conducted two scientific studies in the area.

His first study measured the fear of death levels for 375 undergraduate psychology students from five different Southern California colleges and universities, using the Boyar Fear of Death Scale. Melodyland School of Theology was included in the study. Subjects were placed in Christian and non-Christian groups on the basis of their answers to the question, "Are you a *born again* Christian?"

"No previous study has ever been done where religious faith was identified by people expressing the knowledge that they had been born again by the Spirit of God upon receiving Christ as Saviour," Dr. Cerny says.

His study revealed that the average score of the Christian group was about half as large as the non-Christian, showing significantly less fear of death in the Christian.

"Statistically this difference is so great that it could not have happened by chance more than one in a thousand times," Dr. Cerny explains. "This demonstrates how effective the power of Christian faith is in overcoming the fear of death."[15]

Additional results previously unpublished added a charismatic sample of students found in each of the psychology classes surveyed.

Dr Cerny concludes:

> While the average score of the charismatic sample was not much less than the noncharismatic group, the students surveyed from Melodyland School of Theology expressed significantly less fear of death than any group surveyed.

In his second study Dr. Cerny surveyed 261 secular community college students in their classrooms and divided them into three groups according to their experience of Christian faith:

1. About one fourth of the students identified themselves as *born again* Christians.

2. One half said they were Christians, but didn't know whether they were born again.

3. One fourth said they were non-Christians.

Dr. Cerny says:

> While the first study concerned the *fear* of death, the second identified the *ways* in which the students saw death.

> Those who considered themselves Christians, though not *born again,* had a more positive view of death than the non-Christians. But the outlook of the *born again* group was significantly brighter. It saw death as a door into an afterlife of reward.[16]

A mortician in Stockton, California, makes an observation supporting Dr. Cerny's conclusions:

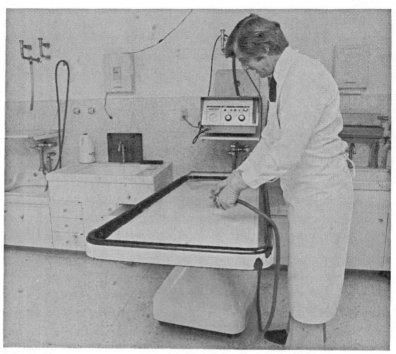

Mortician prepares lab for embalming.

It is less expensive to bury a Christian than a non-Christian. Five minutes, a half hour or a half day before death, the Christian realizes it is the end. The peaceful smile and benign look on his face can't be matched by any embalmer.

The non-Christian has the same experience five minutes before the end and is aware of his being lost. The fear, terror and anguish that comes over his face is so disfiguring that the embalmer has to put in extra time to produce a "natural look."

Dr. Rook says everyone is born with a death instinct, inherent since the fall of man in the Garden of Eden when he was driven from the presence of God. The ego, to preserve itself, will project some of this instinct.

He explains:

> Let's say last Wednesday I was to meet you here at Melody-
> land. But I went to Disneyland instead. Then when I came
> today and saw you sitting here, I'd have feelings of guilt and
> fear.
>
> I would be afraid you were angry because I was thought-
> less, irresponsible and had upset your schedule. I would
> expect retaliation. The fact that you may also have forgotten
> our appointment does not enter my mind.
>
> I would project into you some of my own feelings and
> expect you to somehow harm me. And the ultimate harm is
> death.

The ultimate phobia—the fear of death—has many such facets. A fear of aging, for example, is a projected fear of death because the ultimate end is getting closer; one is helpless to do anything about it, therefore the anxiety. The fear of financial insecurity is another form of this phobia because one often thinks his funds will run out before he dies. The fear of flying, of high places, water or confined spaces—all have their ultimate roots in the fear of death.

Jesus said, "Do not fear those who kill the body, but are unable to kill the soul; but rather fear Him who is able to destroy both soul and body in hell."[17]

It was this rugged courage that enabled the disciples fearlessly to preach, knowing that death was inevitable. When Paul says, "I die daily," he meant he encountered the possibility of death every day he preached. Jesus had no fear of death, for He gladly laid down His life. The Bible says, "God hath not given us the spirit of fear; but of power, and of love, and of a sound mind."[18]

In this chapter we have discussed what happens five minutes after we die, the fallacies of soul sleep and baptism and prayers for the dead. We also have examined some facts about death and how many have overcome their fear of dying.

But how does one prepare for the hereafter? How can I be certain that my financial savings—however large or small—will benefit me in the next life? In the following chapters we will examine these vital questions and show from the Bible how you *can* take your assets with you when you die and *how* you can know where you will spend life in the beyond.

13
Can They Be Brought Back?

Sometimes physicians and ministers have been asked, "Can you bring them back?" The answer is yes. Some are brought back now. But all later. I reiterate that a few do miraculously rise from the dead now. Every person since the beginning of time will come back from the dead in the last day.

If I had the divine prerogative, I would walk through every graveyard and wake up the dead. And if I had the power, I would go through every hospital and heal every patient. But since life and death is determined by God, we must leave it in *His* hands.

David, the psalmist, lamented over his dead son, "I cannot bring him back, but I can go to him" (II Sam. 12:23).

David found comfort in knowing he would see his son in the hereafter.

Death is a sting. Even the mature Christian experiences hurt because loss is painful. I am grieved when people offer such glib phrases as, "All things work together for good," "You'll be all right, honey," "It's for the best," "You'll get over it."

When sorrow strikes, nothing seems good. A cake recipe calls for some bitter ingredients—baking powder, soda, salt, cream of tartar, raw eggs. None of these tastes good alone, but blended together they create a delicacy. Grief is a necessary ingredient in the batter of life. Often the only good that can be explained is God's good. Life has to be interpreted from God's viewpoint.

Let's deal with another of these hollow sayings. "You'll be all right, honey." These words are cold and empty. There *will* be loneliness, but Jesus can fill the emptiness. Someone has said, "A friend is one who steps in when the world steps out." You won't be all right unless you accept a new mode of living.

Here are some of the helps I give in pastoral counseling.

1. *Accept comfort.* Be honest, a person cannot be replaced, either by remarriage or other children. Weep if you wish. Sometimes it helps to empty out the anguish.

2. *Adjust.* Life goes on.

3. *Move,* if necessary. Maybe you can't go on living in the same place because of memories.

4. *Make new friends.* In a world as big as ours, there are millions who've lost someone. You just haven't met them. Accept your new identity. Though you may object, you are now a widow, widower or single. Don't bury yourself in the house. Get out! People aren't going to

look you up. You must find them. Get involved in church, prayer and fellowship groups, community services and civic functions.

5. *Have a hobby.* Get your mind off the past. Live in the present and in the future.

6. *Plan some fun things* for yourself. Play golf. Go fishing. Visit an amusement park. Go to parties. Take a vacation. You may have been shut in with your loved one in an extended illness. It's time to get out.

7. *Face your financial future.* Maybe you *can't* have the same life style as before with your husband's wages. But a simpler life can still be fun. Live within your means, but don't be miserly. Remarriage is one possibility— maybe your financial picture will change.

8. *Don't monopolize your children.* Their needs are different. Although they are the most important part of your life, work and family responsibilities take a toll on their time.

9. *Be positive.* No one enjoys being with a negative person. Though death has dealt a blow, don't become bitter or develop a martyr complex.

10. *Keep on living.* Don't just sit around and mark time until you die. Do something with your life. Accept a new direction.

Even with all of these answers, the scars are there. Let me offer some more sources of strength.

The Comforter Has Come

Jesus said to His followers, "I will not leave you comfortless." Consider the fear and frustration the disciples must have sensed as they waited in the upper room. Their leader had been taken away. Shortly before, the Apostle Peter had denied

that he was one of the Galileans. Now suddenly the Holy
Spirit descended. The profusion of power enabled Peter and
the others to stand fearlessly before an antagonistic mob. The
Holy Spirit not only brought comfort, but made them aggressive in witness.

The Holy Spirit continues to comfort the hearts of God's
people in these days of distress. Let me tell you an experience
where the Spirit brought instant comfort in a crisis.

In the early 1960s when our church had just begun, my
wife and I attended a convention in Phoenix. Some still refer
to that convention as one of the most dynamic ever. It was in
the early stages of the new charismatic thrust, and the meeting
place was packed with pastors and laymen who had come
into the ministry of the Holy Spirit.

One morning during the conference, an offering was being received. A distinguished gentleman in his middle seven-
ties was standing near us. My brother, Ray, who pastors now
in Texas, noted on his convention badge that the man was
Dr. Ed Torsch from Laguna Beach, California.

"Ralph, that man is from your area," Ray nudged. "You
should meet him. Who knows, maybe he is a potential for
your church."

"Hi," I greeted him as he pushed by me in the crowd
returning to the seats. "I'm Ralph Wilkerson, pastor of a new
church that has just started in Anaheim."

"Oh, I'm so glad to meet someone from my area. I came
to this convention to receive the Holy Spirit," he said warmly.

"I would be glad to pray with you," I offered.

"Well, maybe you could come to my room today around
two o'clock," the doctor responded.

"We will be there at two."

At the lunch hour Ray and I sat with two newly Spirit-
filled Church of Christ preachers and another man who was
an interpreter for the United Nations. We told them about

meeting the doctor and of our plans to pray with him that afternoon. All three asked to join us.

When we knocked on the door, Dr. Torsch was awaiting us. After briefly introducing ourselves, I asked the Church of Christ preachers to present the Scriptures on the Holy Spirit. Since they were very new in the ministry of the Spirit, I thought it would be good for them to practice the presentation. Of course, they knew the Bible, but this was a new area of ministry for them.

After a few minutes, the doctor interrupted. "I understand the Scriptures, and I'm ready to receive—NOW."

As he sat in an easy chair with his eyes closed, I slipped over and laid hands on him, saying simply, "Receive the Holy Spirit, *now,* in Jesus' name!" And he burst forth suddenly, but quietly, in a language of the Holy Spirit.

The U.N. interpreter interrupted, "Sir, it is rather unusual. You have spoken the identical phrase in three different languages—French, Italian and Spanish! In every language you said the same thing, 'It has come in time.' "

This phrase made little sense to us at the moment, but later we understood its significance. As the doctor sat in his chair, he did not answer. He smiled, dropped his head and was gone.

We stretched the doctor out on the bed. I gasped, "I think he's dead." When I said the word "dead," those two frightened Church of Christ preachers bolted out the door. I imagine they were afraid of being charged with the doctor's death because of their presence.

I called a famous diplomatic surgeon who also was attending the convention. He instructed me to call the house doctor, since he was licensed in another state.

Waiting for the doctor, we noticed some heart tablets in his shirt pocket. Also lying alongside him on the bed was a Gideon Bible opened to a Psalm. Evidently in preparation to receive the Spirit, he had underlined the words, "I have patiently waited for thee."

It seemed an eternity before the house doctor arrived. We heard footsteps, but they were those of a queenly-appearing older lady, whom I knew intuitively was the doctor's wife.

"Quick, Ray, we must intercept her and prepare her before she comes into the room."

Outside the door was a small, white wrought iron garden seat. We stepped up to her and said, "Would you be seated for a moment? We are the preachers who were praying with your husband today."

"Yes, I know," she nodded.

"But . . . uh . . . uh, he is not doing very well, and we have called the house doctor."

Uncontrollable grief struck at that moment. "He's dead. I know he's dead!" she shrieked.

My brother and I were speechless. No words would come out of our mouths.

Imagine our situation. Inside the room, a dead husband. Outside the door, a sobbing, hysterical wife.

I shot up a silent prayer, "Please, God, give me something to say to this grief-stricken wife."

Then I heard myself saying, "But just before your husband died, he received the gift of the Holy Spirit."

I had hardly blurted out these words when she stopped crying.

"I feel so comforted," she said softly. "Lately Ed had said many times, 'If I can just receive the Holy Spirit, I'll be ready to go home.' "

"We made a trip to a convention in Denver a couple of months ago, thinking he would receive the Spirit. Ed lived for the moment we were to come to this convention in Phoenix, hoping he would receive this time. Just before leaving, Ed had another examination; his heart condition remained unchanged. The doctor told us that he would have to be on constant medication, but he could live indefinitely or go at any moment. We should go on leading a normal life and travel whenever or wherever he wanted to go.

"I particularly quizzed him about the advisability of coming here. But the doctor reassured me it would be all right. Last night I heard Ed up several times, taking his medication."

The funeral service was more victorious than sad. I kept looking at a radiant Mrs. Torsch, a widow with no children. I wondered who would comfort her. Through the next 10 years, before I preached her funeral service, I had the opportunity to witness her unwavering faith and appreciation for the Comforter who had come to abide with her forever.

Relationship with Family and Friends

Often I hear the bereaved remark how much their immediate family means to them during their dark days. At no time is it more important to lean upon those who are close to you. I have noticed that fractured family ties often are healed in the time of death.

Comfort of God's Word

The Holy Bible is the greatest source of strength, for it is unchanging in changing times. Here are some excerpts from a few verses:

> Even though I walk through the valley of the shadow of death, I fear no evil; for Thou art with me . . . whether we are awake or asleep, we may live together with Him. Therefore, encourage one another . . . Let not your heart be troubled. . . . In my Father's house are many dwelling places . . . I go to prepare a place for you (Psa. 23:4; II Thess. 5:10, 11; John 14:1, 2).

Knowing Him

When I was a young preacher, an older man scoffed,

"Don't you know that it's impossible to be assured of your salvation until after death?" What a gamble!

Forty times in one epistle, St. Paul tells us we can know that we are spiritually reborn. No wonder "born again" is now the most popular term in Christian circles.

It was a dark, rainy midnight when Mrs. Wilkerson and I were traveling home after a week of evangelistic services a few years ago. Our car was the first upon the scene of a bad accident. A small sports car had run underneath a large truck, and its top was sheared off. As I stepped out of my automobile, the trucker said, "Mister, you'd better watch it; that car is going to explode."

I looked and saw the mangled form of a man sprawled on the pavement, still conscious.

"Save me," he cried.

Kneeling down beside him, I said, "I'm a preacher, and I want to pray for you."

"We used to go to church, but we got too busy," he mumbled. "Do you think it's too late for me?"

"No. It's never too late. As long as there's breath, there's time. Let's pray right now."

I led the man in the sinner's prayer and asked him if he would receive forgiveness. I saw the peace of God come upon a man who a few seconds before was screaming for help. There was no doubt. God had done a work. He was forgiven.

It's true that you can repent in your last breath and still be forgiven. But I do not recommend deathbed repentance. You have no guarantee that you will be in your right mind to get right with God in the closing moments of your life. You have a life to live for God. And you're the loser by delaying your commitment.

In the next chapter you'll hear a story about Howard Hughes that's never been told to the press. Hughes amassed a great fortune in his lifetime; did he take it with him?

Will we be on the gold standard in Heaven? How can one transfer his assets to the beyond? One of the bypassed subjects that is essential to every Christian is our future rewards. Let's talk about our inheritances.

14
"Howard Hughes— You Can Take It With You"

Each year Melodyland has a New Year's Eve Christian celebration from eleven to midnight. The building is always jammed for this triumphant hour as thousands joyfully watch the old year out and the New Year in.

About three years ago, my wife slipped to the back of the building during the watch night service. The house lights were very low, and she almost bumped into Chico Holiday, a former Las Vegas entertainer, who was standing in the aisle.

"Hey! While I was singing and the lights were up, I couldn't believe my eyes," he whispered. "I looked up and right before me, in walked Karen,[1] H.H. and her two girls. Man, I almost forgot my song, I was so startled."

"See them sitting right over there in section 13? He has on the dark glasses, and that's Karen, wearing a blonde wig and soft shaded glasses for disguise. Her girls are sitting with them," he whispered, nodding in that direction so as not to be conspicuous. "I was able to slip down front after singing and tell Pastor that I had spotted H.H. in the audience."

Minutes before midnight, however, the four of them quietly slipped out as the house lights were down again.

Of course, with Chico and Sally Holiday, Allene and me, the password "H.H." was familiar. H.H. meant Howard Hughes. In telephone conversations, and even privately, we formed the habit of saying Karen and H.H., lest we leaked the name.

Karen, an attractive, wealthy divorcee, came to the Lord through the ministry of Melodyland and became a close friend of Sally's. After being away for awhile, she set up a luncheon appointment with Sally, who is the director of Women's Ministries.

"Sally, I need to talk with someone. I have not one living relative to confide in. I have a big secret to tell you," she began. "I have been dating Howard Hughes, and he has asked me to marry him. . . ."

"Now wait just a minute!" Sally interrupted, fingering her hair nervously. "This is terribly sensational! Nobody ever knows his whereabouts. When we were entertainers in Vegas, and he lived right there on the strip, no one ever saw him. Rumors even floated around that he was dead."

Karen just flashed a faint smile and continued, "Since you and Chico were in show business in Las Vegas before going into the ministry, I thought it would be great for Chico to marry us . . . and Sally, I want you to stand up with me. I have already talked to Howard about this. He has cautiously agreed, if you will do exactly as he orders.

"In discussing our plans, I wanted a simple, private wedding in the pastor's office, with just my two daughters, you and Chico, and possibly Pastor and Mrs. Wilkerson present. But Howard flatly ruled this out. He is adamant about pictures, the press or any publicity invading his privacy.

"He insists on a secret wedding. But he will let Chico perform the ceremony if you will come wherever we are on short notice. Everything is secret, and we will set a code for the time and place to meet us."

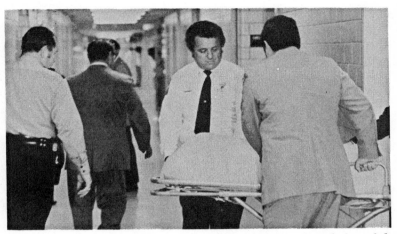

The body of Howard Hughes being wheeled out of pathology lab at Methodist Hospital in Houston, Texas, shortly after his death. (United Press International photo)

Chico kept one of my black wedding robes with him for three weeks so he would be ready. Karen bought an elegant gown for Sally to wear as her only attendant at the wedding.

Through the next several months, over lunch or in the privacy of Sally's living room, she would whisper that she had just met H.H. at a castle in Scotland, or in the Bahamas, Hawaii, or on his yacht. Once she confided that Howard Hughes was cruising off the California coast on some mysterious mission.

Usually when Karen told us where she had been meeting Howard, about a month later the news would report rumors that he had been spotted in those places.

There is no way she could have known his exact whereabouts without being with him. Scotland, the Bahamas and

Howard Hughes' graveside service near downtown Houston. The Rev. Dean Robert Gibson, an Episcopalian, says the benediction. (United Press International photo)

Mexico were accurately established by the press as his residences the last few years. Extensive newspaper articles later told about Hughes attempting to lift a Russian submarine that had sunk off the coast. In our last conversation with Karen, Hughes was in Mexico and desperately ill. He knew he was a dying man. She told us that he had an incurable kidney problem long before it became known to the public. She said he kept a doctor with him at all times. Only a miracle could save his life.

One evening she drove us to San Diego in her new blue, customized car for services. Someone commented on her beautiful gloves as she was driving.

"Oh, yes . . . Howard insists that I wear white gloves all the time," she said matter-of-factly. Later at dinner when she slipped them off, I wondered if she wore them to cover her expensive jewels.

One of the twists that has surfaced in stories about Howard Hughes was his abnormal fear of germs. He required everyone around him to wear white gloves at all times.

Karen told Sally that during her rendezvous with Hughes, they talked about spiritual matters. Karen said she and Hughes had prayed together several times, and she felt certain that he was "born again."

Melodyland, her home church, and particularly the new School of Theology, training ministers for the mainline churches, was mentioned in their conversations. She had hoped that after they married, Melodyland ministries could be included in their will. She made it a point to tell H.H. how he could take it with him by giving to the work of God.

However, almost as mysteriously as Hughes died, Karen, too, vanished. No one has seen or heard from her since his death. At the time of this writing, neither Karen nor the real Howard Hughes' will has surfaced.

One cannot judge another, for only God knows the heart. Who knows? Maybe this eccentric recluse did make his commitment to Christ before he died. If so, did he just get into Heaven? Or does he have rewards?

In the next paragraphs, I will talk about rewards in the beyond.

The Crowns of Christians

To the casual reader of the New Testament, a seeming contradiction over the means and rewards of salvation arises. St. Paul teaches that the sinner is saved by grace alone. Yet, the Christian is rewarded according to his works. Grace pertains to the sin question, but rewards to service.[2]

There are 10 crowns mentioned in the Bible. However, I'll only talk about a few of these rewards. Christian rejoice! God has great plans for you. He not only wants you to have much here, but more over there!

1. *The Soul Winner's Crown*[3]

Often this is called the Crown of Rejoicing. There is no thrill like leading someone to Christ. One of the greatest churches in the world was founded in the early 1900s by a man consumed with soul winning.

This great leader and founder of the Philadelphia Church in Stockholm, Sweden, Pastor Louis Pethrus, left a great impact on the Scandinavian countries.

When I ministered in his former pulpit, someone told me about his concern for souls. At the age of 91 on the deathbed of this pastor, one of the leading merchants of the city came to see him.

In a faint whisper the aged minister said to this merchant, "I have visited with you many times in your store. You have respected me and have been my friend. I am about to go beyond. Shall I tell them that you, too, will be coming later?"

The businessman gave his heart to Christ moments before the pastor slipped on to his reward. There is no question that Pastor Pethrus will inherit the Soul Winner's Crown of Rejoicing.

If one person won one, and the two won two, and the four won four and this multiplication process continued, the entire world would be won to Christ in 33 days.

2. *The Overcomer's Crown*[4]

This is sometimes called the Crown of Life. Bible scholars link it with the Crown of Righteousness laid up for Paul and those who have perservered for the heavenly prize.

This incorruptible crown[5] will be awarded to committed Christians who have laid their lives down for Christ.

Tradition gives us one of the last scenes of Paul's life.

As he trudged wearily down the cobbled streets of Rome on his way to the chopping block, he was chained to two Roman soldiers, one on each wrist. They passed the famous statue of Nero, which was inscribed with the word, "Conqueror." One of the soldiers looked at Paul and said, "With your great abilities and power to persuade men, you too could have had a statue with the word 'Conqueror' on it."

Paul did not reply, but hung his head. The soldiers heard him mumbling these words as his voice faded away, "More than conquerors. More than conquerors. More than conquerors. . . ."

Crowns are never given as souvenirs. They are won.[6]

3. *The Crown of Glory*[7]

I like to think of this crown as one that is reserved for pastors or Christian workers. Some believe preachers should be poor as a mark of humility. The Bible doesn't support this.

Abraham, Isaac, Jacob and Joseph, who were prophets of God, were some of the wealthiest men of the Old Testament, yet they were the closest to God and the most humble in commitment.

The story of Job is a classic. Though he was stripped to poverty, material blessings were lavished upon him. According to the final account in the Book of Job, he had more in the end than in the beginning.

Jesus spoke to His disciples about the rewards of the Christian life, and He promised a hundred fold now in houses, brethren, sisters, mothers, children and lands, and in the world to come eternal life.[8]

Regardless of financial status, the faithful servant will receive this shepherd's crown. An added reward is that we will reign with Christ upon heavenly thrones.

4. *The Crown of Life*[9]

This crown belongs to the martyr. In the Book of

Revelation, the emphasis of faithfulness is not until death, but *unto* death. When charged before magistrates or thrown to the lions, the early Christians kept their commitment *unto* death.

Why Should I Want to Wear a Crown?

We have discussed different crowns, but what is the significance of these inheritances?

First, authority is symbolized by a crown. It is a vestment of power. The crowning is the official coronation of a king.

Second, the crown is a symbol of purity in works. God judges our motive. He knows the intent of our heart.

Third, a crown is the only tangible tribute we can offer to Him. It is said that as one of the queens of England was crowned, she cried, "Oh, that I may lay my crown at the feet of the King of Kings." Giving the crown is another expression of worship in the hereafter.

Fourth, crowns will be converted to positions. Our works will be transferred beyond as approvals in the Bank of Heaven. The highest form of reward is when one stands or sits by Jesus, realizing that one has pleased the Lord in this life.

The most desired place is to be next to His throne. The mother of James and John came to Jesus, requesting this special place. You can't blame the woman for looking out for her sons. She wanted one son to sit on the right side of Jesus and the other on the left side in His heavenly kingdom. Jesus did not rebuke her. He simply said it was not His right to determine this. It was His Father's decision.

What Are the Future Finances?

What is the currency of the beyond? How would you spend money if you had all you wished for? What value is it to have a crown?

Money is nothing but a means of carrying on commerce.

The Indians used to exchange arrowheads. The early trappers traded with furs. In Africa, barter beads were used for money.

These could not be the medium of exchange in Heaven:

> Gold — because the streets are constructed of it.
> Pearl — because the gates are made of it.
> Precious gems — because the wall of the celestial city is solid with them.
> Rare jewels — because the wall's foundation is studded with them.

If your bank account, large or small, could be carried with you as you leave this world and go into the next, *how would you spend it?*

It couldn't be for clothing, for you will wear the garments of God.

It couldn't be for food, for you will eat the fruit of the tree of life.

It couldn't be for beverages, for you will drink of the water of life.

It couldn't be for doctor bills, for there is no sickness there.

It couldn't be for a new house, for Jesus has prepared many mansions for you.

It couldn't be for a much-needed vacation, for you will have pleasures evermore.

So your finances are of no value in the beyond until they have been converted into the currency of glory—rewards and crowns.

The greatest things in this life cannot be bought with money—love, health, happiness, friends, peace. In this world happiness can become hollow. Love looks for a lover. Health can be hopeless when money for doctor bills runs out. Friends can turn out to be fair-weather ones. Peace can shift to panic. But in the next life, these are everlasting and perfect, and they are the Christian's inheritance.

When Do You Cash These Dividends?

God advances some of the interest on these benefits now, while the principal is still intact in the Bank of Heaven. Some of the assets of eternity—love, peace, joy—are ours to enjoy here.

Far too much emphasis has been placed upon the sweet bye and bye when we receive our eternal rewards. Heaven begins here. Heaven can be in your heart and home now, as well as in the hereafter.

The test of true works is whether we have built on Jesus Christ. Many self-centered people in churches will be shocked to find they have nothing to offer Jesus. Their services have been given for recognition. They already have their reward. Others give to have their name appear on stained glass windows or to have buildings named after them. If given only for attention and self-satisfaction, their reward is now. But given to the glory of God, their reward is eternal.

The Bible teaches that all Christians will some day stand before the Lord—not to be judged as to whether they enter, but to be examined for their works. The second appearance of Christians will be at a wedding feast in Heaven. At that time it is most important that we have on our wedding apparel[10] to be properly attired. We dare not enter His presence without adequate preparation.

When we were in our twenties, we conducted evangelistic meetings in a fashionable Southern California beach town. The church building was not completed, and temporary living quarters for the pastor and his guests had been set up in the basement.

About two in the morning I heard a light rap on the door. "Ralph, get up quick. We have a prowler. Let's catch him." Just then there was a loud crunch on the graveled parking lot as a temporary power pole toppled. I heard the pastor's steps racing up the stairs.

I could picture the pastor being clubbed by a burglar. I

darted out the door. About a block away we found the in-truder hidden in some bushes. Arm in arm, one on either side, we walked him back to the church to call the police. Suddenly I noticed cars slowing down and passengers peering at us very curiously. Only then did I realize I had forgotten to put on my trousers. I was wearing my underwear—a tee shirt and boxer shorts.

The embarrassment I felt that night must be typical of the regret that one will sense when he stands before God. Some will stand stripped before God because they made no prepara-tion for the wedding except salvation.

How Can I Transfer My Account Beyond?

Someone suggested to me, "Give it to one who's on his way to Heaven." How do you get your assets from Planet Earth to Planet Heaven?

It's a known fact that when one gives to others in this life in the name of Jesus, it's like putting it in the Lord's hand in Heaven. Jesus said, "Whosoever shall give you a cup of water to drink in my name, because ye belong to Christ, verily I say unto you, he shall not lose his reward."[11]

One of my classmates, Ernie Reb, a missionary in the Philippines, has built several hundred churches. One of his first ministries was in Manila in a huge evangelistic center numbering in the thousands.

Now God has taken him to the "bush." One day I asked him, "How do you introduce people to Christ who are totally pagan?" He told me an intriguing story:

I built my house on a path where two different tribes pass regularly. It's a great distance to the springs or rivers where we located our house.

I tried preaching to the natives, but with their stoic look there was no response. I attempted to find out why other missionaries had failed, for not one of the tribes had become Christian.

I stopped preaching and started practicing Christ. The next day an elderly man from one of the distant tribes came by. From our deep well, I took a cup of cold water and simply said, "I give you this drink in Jesus' name."

Ralph, it was only a matter of days until more of the tribesmen came by to drink my water. I always told them that it was Christ's water. And He offered living water."

Soon they were not merely drinking natural water, but they were drinking the water of life—Christ. Now the entire tribe is coming to Jesus.

Money Has No Value Until Spent

Currency is nothing but a means of carrying on commerce. If you place your money in a bank and never spend it, you will receive no value. The biblical proof that one can take his resources with him is the verse, "Lay up for *yourselves* treasures in heaven. . . ."[12] The phrase *yourselves* places it on a personal basis. In the day of the diminishing dollar, it's so important to make wise investments. Huge fortunes are meaningless unless one plans for their distribution while he is yet living.

Think what Howard Hughes could have done with his huge fortune. Had he carefully planned his estate and utilized it for the evangelizing of the world, he would have treasures in Heaven. Two others of the world's wealthiest men died recently. J. Paul Getty and Aristotle Onassis. As far as I know, neither of them gave directly to God's work. Perhaps they had planned to give to God, but waited too late.

Did you know that 70 percent of all adults die without a last will and testament? And that 80 percent of those who do have a last will and testament have not taken advantage of approved tax savings plans to preserve their estates?

There are many excellent Christian organizations that have departments set up for the purpose of assisting you, should you wish their help.[13]

Sometimes one of our only means of service to God is being stewards of what God has given. Often a successful person doesn't have time or talents for other ministries, so his greatest investment in the hereafter is a wise use of his money.

Let me also emphasize that it is not merely the amount that one gives, but attitude. Jesus commended the widow who gave her two mites. I often tell people I know where Jesus sat when He went to church. This story says *He sat over by the treasury* (Mark 12:41).

In Hebrews 6:10 God says He is not unrighteous to forget one's work, his love and his service to His saints.

In the conclusion, may I again emphasize the words of Jesus, ". . . for where your treasure is, there will your heart be also."[14]

In the coming chapter you will experience the ultimate trip into the celestial city, and I will discuss some of the questions that have intrigued many of us. Do those in Heaven know what's going on here? Do we know the exact size of Heaven? Are the descriptions symbolic or literal? Will we know each other there? What do people do when they get there? What are the chief attractions of this planet called Heaven in the outer galaxy?

15
The Ultimate Trip

Maj. A. G. Nikolayev and Lt. Col. Popovich, the Russian cosmonauts who orbited the Earth in outer space, reported they sighted no trace of anything resembling Heaven. It must not exist, they concluded.

In contrast, at least three American astronauts saw enough of outer space to propel them into full-time Christian service. Their belief in Heaven is evidenced by their new mission in life.

Undeniably, no cosmonaut or astronaut has seen Heaven. But positioned somewhere in outer space, beyond the vast reaches of man, is this celestial planet, the headquarters of eternity.

In this book we are using the terms *New Jerusalem* and *Heaven* interchangeably. However, the Bible does differentiate between the two. The New Jerusalem is a city on the planet Heaven, just as Los Angeles is a city on planet Earth.

Very little description is given of the planet Heaven. But extensive detail is given by John the Revelator about the city New Jerusalem. John saw that holy city *coming down out of Heaven* from God to Earth.[1]

Substantial evidence indicates that many cities exist on planet Heaven. Though none of these metropolises match the magnificence or the size of the New Jerusalem; their splendor defies description.

Most people have a concept that because Heaven is exclusive, it must be small. Absolutely untrue! Imagine the immensity of a city stretching 1,500 miles. A city so immense it would overlap the geographical boundaries of the entire western United States, north to south. And this is only *one* city on planet Heaven.

Add to this the vast regions of paradise plains carpeted with miles of exotic flowers and forests against the backdrop of distant majestic mountains. Proportionately, planet Heaven would prove to be many times larger than Earth. We already know that many other planets in our solar system are larger than Earth—Mars, Jupiter, Saturn, Uranus. . . . Our sun is approximately 109 times the size of Earth, and literally millions of planets in the universe are larger than the sun.

Many who have gone beyond report visiting these far-flung cities and park-like gardens. I accept their reports as true, not only because they have seen them, but because they harmonize with biblical accounts.

Astronaut Jim Irwin has appeared at Melodyland several times. Once he told me that two things in space travel are most staggering: the smallness of the Earth and the vastness of the universe. Today much is known about the planetary heaven that was unknown a few decades ago.

Heaven is not some mere abstract fancy of man, but is as real as any city on Earth, such as San Francisco, London

To MELODYLAND CHRISTIAN CENTER WITH MY LOVE IN CHRIST, Jim Irwin APOLLO 15
5 SEPT 1979

HIGH FLIGHT FOUNDATION

(NASA photo)

or Tokyo. One interpretation of Heaven isolates it on a spiritual plane. Thus, Heaven would be the ultimate level in
degrees of spiritual development. Jesus said, "I go to prepare
a *place* for you."[2] Though invisible to us, it is a literal,
physical place.

The Bible speaks of three heavens—the aerial heavens,
the planetary heavens and the Heavens of heavens.

The first, where the rain clouds are formed and the birds
fly, is where Satan has his throne, for the Bible calls him the
prince and power of the air.

The second is where the sun, moon and stars declare the
glory of God as the psalmist affirms. It is through this realm
that American astronaut Irwin flew nearly a half million miles
to walk on the moon . . . and found God!

The third, God's residence and throne, is ultimately where
all believers will live.

Speculation as to the location of Heaven has long been a
question of man. The Bible always refers to Heaven as being
up. A cloud received Jesus *up* out of the disciples' sight.
When Christ comes for His church, we also will be caught *up*.
Elijah was caught *up* in a whirlwind.

A country preacher once preached on how Elijah was
bodily transported to Heaven. After the service a skeptic
cornered him.

"Sir, I question the validity of your sermon tonight on
Elijah's bodily translation," he challenged. "Don't you know
that the outer atmosphere is so rarified that the prophet would
have turned into a block of ice as he traveled through space?"

"Shucks, the Lord took care of that, Mister," the preacher
drawled. "He sent along a chariot of fire and hosses of fire
to keep Elijah warm all the way from here to glory."

"Furthermore, as Elijah raced through space, he could
have come close to a fiery constellation or starry body and
have burned to death," the scoffer continued.

"Oh, that was taken care of, too. . . . The Lord sent along

a whirlwind to air-condition Elijah's chariot and to keep him cool on his trip up yonder."

But which way is *up*? Following the logic that *up* is at right angles to the Earth's surface wherever one may be standing, then *up* is nowhere in particular but everywhere in general. Because the world is round, people on the other side of the globe looking up would face a different direction.

Back to the question, exactly *where* is Heaven located?

By the process of elimination in Psalm 75:6, north is established as the location of Heaven. Isaiah, the oratorical seer, concurs with this point. In the 14th chapter, he refers to the position of God's throne as "in the recesses of the north."

Scientists have observed that the solar system is rapidly speeding toward the north. That magnetic attraction pulling us upward is undoubtedly the magnetism of Heaven's forces. Job 26:7 says, "He stretcheth out the north over the empty place, and hangeth the earth upon nothing."

Two of our dearest friends, Oral and Evelyn Roberts, lost their daughter and son-in-law recently in a fatal airplane crash. Through the years the Roberts have effectively ministered to the masses in healing and comfort. Even in their hour of deep sorrow, they went on national television to tell their listeners about it. Many have remarked that this magnificent demonstration of courage has brought healing to their hurts.

Evelyn later wrote us that the resurrection and Heaven have become so much more real to them since their children have gone there.

Why Will Heaven Be Heaven?

The biblical description of Heaven is an array of positives and negatives. Heaven will be Heaven because of what and who will be there, and because of what and who will not be there.

The greatest attraction of Heaven will be Jesus. When

John the Revelator was shown the "things which should be hereafter," he gazed first not at "things" but on the "One who sat upon the throne."

When the martyr, Stephen, was being stoned, he looked into Heaven and saw ". . . *Jesus* standing at the right hand of God."[3]

Years ago I titled an evangelistic sermon *Something Better Than Heaven and Worse Than Hell.* I asked the question, "What could be better than Heaven, yet worse than Hell?" One of the teenagers whispered, "Marriage!" Maybe in some marriages this could be true, but that wasn't my subject. Seeing the face of Jesus will be *better than Heaven* for some, but for others it will be *worse than Hell.*

The Bible says, "The face of the Lord is against them that do evil."[4] Revelation 6:16 predicts, "They will cry for the rocks and the mountains to hide them from the face of Him who sitteth upon the throne." More terrifying to some than the sun fading, the moon bleeding, graves bursting, the world crumbling and the elements burning, will be the sight of the face of Jesus on judgment day.

Inhabitants of Eternity

Immediately after seeing the face of Jesus, our second attraction will not be the streets of glassy gold nor the gates of glory, but our dearest loved ones.

When one ponders the facts of eternity, the paramount questions are about marriage in Heaven and parental relationships with children.

"Whose wife is she in eternity?" asked the sad Sadducees, attempting to discount the resurrection.[5] This zinger they threw at Jesus concerned a woman who outlived her husband and six of his brothers whom she married according to Mosaic law. If all seven brothers go to Heaven, which one would be her husband? the Sadducees quizzed. (I thought—Who would want her? If she outlived all seven brothers, something must be wrong.) When Jesus answered this trick question, He gave us a peek into eternity. In Heaven marriage ties do not exist.

I *do* believe that we will retain an awareness of our relationship as husband and wife or father and mother. However, there will be no physical union, for there is no need of procreation in Heaven.

Will Infants Be Babies in Heaven?

The supreme joy of this world is our children, the reproductions of us. We revel in sentimental memories of our children crawling on the floor, playing in the yard and being tucked into bed at night. The absence of children in Heaven would be disappointing.

Jesus said, "Suffer little children, and forbid them not, to come unto me; for of such is the kingdom of heaven."[6] Our heavenly Father has included children in the society of Heaven.

Some theorize that babies will remain infants in Heaven. Perhaps some of this might be attributed to artists' paintings of tiny cherubim in ethereal settings. Though the art is beautiful and inspiring, don't confuse cherubim with children. Cherubim are angels, and a cherub is not a baby angel. It's merely the singular of the plural, cherubim.

One of the first persons I bounced this question to was Chico Holiday. He and his wife, Sally, former entertainers in Reno and on the strip in Las Vegas, have a family of eight adorable children. I figured Chico should be pretty qualified to answer my question. "Chico, when we get to Heaven, do you think that babies will remain as infants?" I asked soberly.

"Good heavens! I hope not!" he exclaimed. "Do you have any idea what it would be like, changing diapers, mixing formulas and listening to 'goo . . . goo . . . goo' throughout eternity? What a waste of creative power to create a permanent infant."

It would seem that there is no established chronological

age in Heaven. Children and old people will appear to be one
ideal age. One of my Bible professors, Dr. W. B. McCafferty,
in supporting this theory, used the logic that Adam and Eve
were not created as infants, but were full grown. He also
contended that old age is a mark of sickness and sin. So older
people will be returned to youth and infants matured to
adults to create a uniform society.

Another conclusion is that whatever God prepares in
Heaven is perfect and complete. There will be no imperfections materially, spiritually or intellectually.

One cannot be dogmatic in conclusions about some of
these areas of eternity, for all has not been revealed. We have
never seen gold as transluscent as glass, a pearl so huge it
can make a gate 100 miles wide, nor trees bearing 12 different
kind of fruit. Let's keep an open mind and be free for con-
jecture, if it does not conflict with Bible teaching on the
hereafter.

Will We Know Each Other There?

The wife of the quaint Scottish preacher, John Evans,
once asked her husband, "Do you think we will know each
other when we get to Heaven?" To which he replied, "My
dear, do you believe we will be bigger fools in Heaven than
we are here?"

With an emphatic yes, I must say that recognition and
personality extend beyond the grave.

In Heaven recognition will not be based on physical appearance, for we will know people we never met in this life.
St. Paul said, "We only know in part, but then shall I know
even as I am known."[7] This implies a divine sense of knowing.

On the mount of transfiguration, three of Jesus' closest
disciples identified Moses and Elijah, whom they had never
seen. Jesus tells us in Matthew 8:11 that we will sit down with
the patriarchs in Heaven. To enjoy their company, we must
know who they are.

Personality projects beyond the grave. The most identifying thing about you is not your size or color of hair and eyes, but your spirit. In Heaven recognition will not be based on physical appearance.

The Emmaus disciples did not recognize the resurrected Lord. Evidently His physical form had altered. But when Jesus prayed, instantly they knew Him. Your love, sense of humor, emotional reactions and even the look in your eyes reveal individual personality.

They See Us Here

St. Paul in Hebrews depicts the scene of the amphitheater in his day. Perhaps this was the setting where he fought with the ferocious beast in the arena. Far in the distance among the thousands of spectators, the apostle caught a glimpse of his fellow saints praying for him. Though they were unable to communicate, he sensed a burst of inspiration, just knowing his friends were supporting him.

The spectators in the amphitheater of Heaven are the believers who have traveled beyond. Paul says we are encompassed about with so great a cloud of witnesses.[8] This cheering throng can see our everyday activities. Enormous encouragement comes just knowing that I have Heaven's forces battling with me in the struggles of this world.

When in the Philippian jail, Paul undoubtedly drew courage from this. On the Isle of Patmos John the Revelator had a vision of Heaven. He saw a host that no man could number, praising God around the throne.

Our Lord also taught us that there is joy in the presence of the angels of God over one sinner who repents.[9] Note that it does not state that the angels are rejoicing, although I'm sure they are. But it says there's joy *in* the presence of the angels.

Who is the happy crowd? I believe it's a mother who has prayed for her son, a pastor who prayed for spiritual renewal in his church, or possibly a missionary who died on some distant shore, proclaiming Christ without ever seeing the results. One day those prayers were answered. Yes, answered after they had departed to the land beyond. And now Heaven is rejoicing.

Happiness in Heaven with Loved Ones in Hell

Let me pose a new problem for you. How could there possibly be happiness in Heaven, knowing that you have loved ones who are lost? The answer is in the Bible. Job 24:20 says, the sinner ". . . shall no more be remembered." Again, Isaiah 65:11 declares, ". . . the former shall not be remembered nor come into mind." God teaches us that He will erase the memory of them forever. In Ecclesiastes 9:6 He says, ". . . the memory of them is forgotten." In Heaven the believer will have a panoramic view of the Earth from God's standpoint.

Heaven Is Large Enough for Everyone

It takes people to make a city. In the New Jerusalem, John saw ". . . a multitude which no man could number."[10]

Critics have charged that Heaven would not be large enough for all those who go beyond. We know for certain that Heaven is a perfect cube 1,500 miles high and 1,500 miles square.[11] Imagine a city that would stretch all the way from Maine to Florida! Since some of the metropolises on Earth cover large geographical areas, this is not hard to conceive. The largest city in the world in area is Mount Isa in Queensland, Australia, with 15,822 square miles. Cities with less territory most often have larger populations—Los Angeles, New York, London, Tokyo. Of these, New York is the largest, with more than 16 million people. How many will inhabit the heavenly city? Perhaps billions. To accommodate those masses, literalists have suggested some of the following staggering propositions:

1. Putting each street a mile apart, there would be eight million streets, each 1,500 miles long.
2. If the city were divided into rooms one mile in length, one mile wide, and one mile high, it would contain three billion, 375 million rooms.
3. Another interesting calculation is to multiply 1,500 miles by 1,500 miles; you would have 2,250,000 square miles. There are 640 acres to a square mile, so you have one billion, 440 million acres in the city.

God has made the gates of Heaven large enough to accommodate all who wish to come through. Someone said that for these gates to be properly proportioned to the wall of the city, they would have to be at least 100 miles from pillar to post. It's astonishing to note that each gate will be made of a single snow white pearl. The largest pearl known is about the size of a man's fist.

The Activities of Heaven

Is Heaven a place of inactivity where people wear crowns and while away their time, strumming on harps? I hardly think so.

Heaven will be a place of pleasure. Celebrations, singing, fellowship, sightseeing, dining on delectable fruits, wonder at the attractions of the city and reveling in the luxuries of eternity will be just a few of our heavenly delights.

Not only will we have pleasure, but purpose in eternity. The Bible talks about going in and out of the gates. Business will be conducted in Heaven. But it will be God's business. If there were no movement in and out of the city, its gates would be unnecessary.

Rest will not be needed, for our bodies will not tire. The Bible says there is no night there.

Heaven is a place of praise. The highest engagement is the worship around the throne. The Bible says, "The angels always do behold his face."

Both worship and praise will be offered as gifts unto God. The difference between the two is that worship is bowing down and praise is lifting up. One will bow down in reverence to Him and will lift up holy hands in praise for His redemptive powers. Certainly if one does not enjoy worship and praise here on planet Earth, he would feel most uncomfortable in Heaven!

How Long Will One Be in Heaven?

Eternity is the lifetime of God. Just as space and distance is relative, so is time. The Bible says, "Time shall be no more." Forever one will be in the presence of God.

We will leave Heaven to rule on this Earth with Jesus for 1,000 years. There will be a complete renovation of this Earth and only righteousness will prevail. Then we will have Heaven on Earth. But as the most popular hymn of the church, *Amazing Grace,* goes, "When we've been there ten thousand years . . . we've just begun."

When man strives to stay on this planet forever, it's most difficult to understand the words of St. Paul that Heaven is far better than this world. [12]

Heaven is far better *socially*. The choice aristocracy will be there. A select company from every tribe, tongue and nation will be represented.

Heaven will be far better *physically* because there will be no pain. Imagine a place with no hospitals, no suffering or no disfigured bodies.

Heaven will be far better *financially,* for there will never be a need in Heaven. In times of worldwide economic crunch many struggle for mere existence.

Last, Heaven will be far better *environmentally,* for there will be no smog to obscure the skies. None of the devastations known to man, such as tornadoes, earthquakes and flood, will be evident.

The space program has been most concerned, not only about the contamination of its space vehicles, but with pollution being carried from planet to planet.

Heaven is a pure place. And God will not have it contaminated by germs, disease or anything unclean. Just as all space vehicles, clothing and equipment must be decontaminated in space preparations, so must we be spotlessly cleansed.

Pollution is a type of sin that spoils a perfect environment. Who produces sin? As long as Satan exists, evil will exist. The absence of Satan and sin in itself will insure that Heaven will be Heaven.

The last two chapters of the Bible emphasize the negatives with seven "no mores." Imagine a place with no more sea, no more death, no more pain, no temple, no need of sun, no night and no candle, as John the Revelator exclaims. Man does not have adequate vocabulary to describe the glories of the hereafter. Paul says:

> Eye hath not seen, nor ear heard, neither have entered into the heart of man, the things which God hath prepared for them that love him. But God hath revealed them unto us by his Spirit.[13]

Imagine peacefully landing on a shore and realizing it's Heaven's shore; drinking pure water, and learning it flows from the river of life; eating delectable fruit and finding that it's from the tree of life; inhaling clean air, and knowing it's celestial air; experiencing thrills and sensing immortality; feeling a touch and taking hold of the nail-scarred hand; of waking up and discovering we've made the ultimate trip!

Chapter 1

[1]Ford, Marvin, "Thirty Minutes in Heaven," *Full Gospel Business Men's Voice* (Full Gospel Business Men's Fellowship International, P. O. Box 5050, Costa Mesa, Calif.), October 1976, pp. 7-9, 12, 13.

[2]*Family Circle*, November 1976. Charles Panati is a physicist who taught at Columbia University and has been a science editor at *Newsweek* for several years.

[3]Kübler-Ross, Elizabeth, M.D., *Questions and Answers on Death and Dying* (Macmillan Publishing Co. Inc., 866 Third Ave., New York, N.Y., 1974), introduction p. xii.

Chapter 2

[1]Esses, Michael, *Michael, Michael, Why Do You Hate Me?* (Logos International, Plainfield, N.J., 1973), pp. 113-119.

Chapter 3

[1]Michael Esses at this time was head of the Melodyland School of the Bible. He had been sick with his heart and other complications, and many times I had prayed for him. He has since become author of three best-selling books, *Michael, Michael, Why Do You Hate Me?*, *The Phenomenon of Obedience* and *Jesus in Genesis*. He is a former rabbi.

[2]Pp. 140-149.

[3]John 11:25; Luke 10:19; John 14:12, 13; Matt. 18:20.

Chapter 4

[1]Prov. 16:25.

[2]Matt. 6:33.

[3]John 3:8.

[4]I Corinthians 12.

Chapter 5

[1]John 2:9-11.

[2]Matt. 14:29.

[3]Acts 8:39, 40.

[4]Luke 7:14, 15.

[5]Tari, Mel as told to Cliff Dudley, *Like a Mighty Wind* (Creation House, 499 Gunderson Drive, Carol Stream, Ill., 1971), pp. 66-78. While I'm aware there are many skeptics who do not accept Mel Tari's reports, I believe him to be an honest, sincere man of God. He has spoken at Melodyland. However, I must let his accounts vouch for themselves, for I have not personally examined the cases he tells about.

[6]Dr. Madrigal is a doctor of internal medicine at Universidad Nacional Autonoma de Mexico.

[7]I Cor. 14:15.

[8]This incident is corroborated by Luisa in testimony translated from Spanish: "When Margarita Rodriguez die, Felipe and Doris, with their prayers, made Margarita return to life for one day. Next day at 12 noon she die again. I, Luisa and my husband, as witnesses, saw it. She was stiff, and she was not breathing. But with their prayers asking Jesus, my grandmother had one more day of life." (Written by Luisa, wife of Juan Castillo; signed by Juan Castillo.)

Chapter 7

[1]"The Exorcist," *Newsweek*, Sept. 10, 1973, p. 31. Also, "A Matter of Faith," *Time*, Sept. 10, 1973, p. 76.

[2]Parker, Larry, "I Watched My Son Die," *Acts* magazine (World Missionary Assistance Plan, Burbank, Calif.), Vol. 5, No. 3, pp. 18-20.

[3]"Prayer Didn't Revive Corpse," *Daily News Tribune,* Fullerton, Calif., Jan. 19, 1976, p. A6.

[4]"Prayers Fail to Raise Body in 17H," *Los Angeles Times,* Dec. 10, 1976, Part 1, p. 4.

Chapter 8

[1]*The Sun-Herald,* March 14, 1976.

[2]Mr. Isaacs attached a statement to his letter from a specialist at the Southern California Permanente Medical Group, Alan R. Cantwell Jr., M.D., which reads: "This patient has had a very unusual healing of a condition for which he had no direct medical treatment. I am very surprised at his healing."

[3]Charismatic clinics are held the second week of August each year.

[4]Interview in *Christianity Today,* July 20, 1973.

[5]James 5:15, 16.

[6]III John 2.

[7]Psalm 66:18.

[8]Rom. 8:11.

[9]Rom. 10:9, 10.

[10]III John 2.

[11]*Questions and Answers on Death and Dying,* pp. 19, 20.

Chapter 9

[1]Rev. 11:9. By satellite television every part of the world will simultaneously view this event in Jerusalem.

[2]Revelation 6, 7, 8.

[3]Acts 10:36.

[4]Rev. 11:5.

[5]Rev. 11:7.

[6]Rev. 11:9.

[7]Rev. 11:10.

[8]Rev. 11:2.

[9]Rev. 11:11.

10Rev. 11:13.
11Matt. 12:40. See also Luke 11:29, 30
12Luke 8:55.
13II Cor. 12:2-5.
14II Cor. 5:8, Eph. 4:8-12; Rev. 6:9.
15Adapted from John 5:25.
16See Luke 24:13-31.
17Luke 24:38-40.
18John 20:26.
19John 20:14-16; Luke 24:14-16, 31.
20I Thess. 4:13-18; 5:1-3.
21Rev. 20:5-15.
22John 14:6.
23I Cor. 15:22.
24Heb. 9:27.
25Roberts, Rev. Alexander, D.D., and Donaldson, James, LL.D., eds., revised by A. Cleveland Cox, D.D., "Irenaeus Against Heresies," *The Apostolic Fathers with Justin Martyr and Irenaeus,* American edition (Wm. B. Eerdmans Publishing Company, Grand Rapids, Mich., 1973), Book 2, p. 409.

Chapter 10
1Rev. 1:10; Mark 16:2; Luke 24:1.
2Acts 20:7.
3Heb. 7:25.
4John 14:16.
5Rom. 6:4, 5.
6Acts 4:33.
7Eph. 2:1.
8I Cor. 6:14.
9Rom. 10:9, 10.
10Luke 24:37-39 *King James* Version.
11Isaiah 53; Ezekiel 37.

Chapter 11
1"Cruel Questions: Case of K.A. Quinlan, M. Shields and Others," *Newsweek,* Sept. 29, 1975, p. 76.
2Berger, P.F., and Berger, C.A., "Death on Demand," *Commonweal,* Dec. 5, 1975, p. 586.
3Vaux, Kenneth, "Beyond This Place: Moral-Spiritual Reflections on the Quinlan Case," *The Christian Century,* Jan. 21, 1976, p. 45.
4Lyons, Cathie, "The Quinlan Decision," *Christianity and Crisis,* May 10, 1976, p. 103.
5Oden, Thomas C., "A Cautious View of Treatment Termination," *The Christian Century,* Jan. 21, 1976, p. 40.
6Rice, B., "A Time to Live and a Time to Die—How Physicians Feel," *Phychology Today,* September 1974, pp. 29, 30.

[7]Cant, G., "Deciding when Death Is Better than Life," *Time* essay, *Time*, July 13, 1973, p. 37.

[8]*Ibid.*

[9]Psa. 37:36, 37.

[10]"Medical Ethics—Who Decides the Life and Death Issues?" *U.S. News & World Report,* June 16, 1975, p. 63.

[11]*Ibid.*

[12]Wiley, John P. Jr., "Immortality and the Freezing of Human Bodies," *Natural History,* December 1971, p. 13.

Chapter 12

[1]Morison, Robert S., "Dying," *Scientific American,* September 1973, p. 55.

[2]*Ibid.,* p. 56.

[3]Kübler-Ross, *op. cit.,* pp. 1, 2.

[4]Morison, *op. cit.,* p. 55.

[5]*Ibid.*

[6]"Why should you die before your time?" Eccl. 7:17.

[7]Rev. 20:14.

[8]I Thess. 5:23; Gen. 2:7.

[9]Gen. 3:19.

[10]Winter, David, *Hereafter, What Happens After Death?* (Harold Shaw, Publisher, Wheaton, Ill., 1973), p. 26.

[11]*Ibid.,* pp. 26, 27, 29, 31.

[12]Eccl. 12:7.

[13]Based on an interview with Dr. Walter Martin, a noted authority on this subject and a professor at Melodyland.

[14]II Cor. 13:1; Matt. 18:16; Deut. 19:14.

[15]Cerny, Leonard, J., interview. Facts supported by his article in "Christian Faith and Thanataphobia," *Journal of Psychology and Theology,* 1975, Vol. 3, pp. 202-209.

[16]An indepth analysis of this study is documented in his unpublished doctoral dissertation, "Death Perspectives and Religious Orientation as a Function of Christian Faith" (Rosemead Graduate School of Psychology, 1977).

[17]Matt. 10:28.

[18]II Tim. 1:7. See also Heb. 2:14, 15.

Chapter 14

[1]Karen is not her real name. We have changed it to protect the person in this true story.

[2]Eph. 2:8, 9; James 2:17.

[3]I Thess. 2:19.

[4]James 1:12; II Tim. 4:8.

[5]I Cor. 9:25.

[6]I Cor. 9:24.

[7]I Pet. 5:4.

[8]Mark 10:30.

[9]James 1:3; Rev. 2:10.
[10]Rev. 19:9.
[11]Mark 9:41.
[12]Matt. 6:20.
[13]Few Christians realize that their ability to present an effective Christian witness does not end when they go beyond. Each Christian should have a *Christian will* including a written personal testimony to God's grace. This witness is inevitably read by attorneys, legal secretaries, judges, trust officers and family members who may not know Christ. Our personal witness for Christ can continue through our last will and testament.

Melodyland has an Estate Planning and Trust Department, which provides confidential help with these matters. There is no obligation to include Melodyland in your will or other plans. We believe we can show you how to conserve your estate, continue your witness, and at the same time give to God's work. Write: Estate Planning Department, Melodyland, P. O. Box 6000, Anaheim, California 92806.
[14]Luke 12:33, 34.

Chapter 15
[1]Rev. 21:2.
[2]John 14:2.
[3]Acts 7:55.
[4]I Pet. 3:12.
[5]Matt. 22:23-30.
[6]Matt. 19:14.
[7]I Cor. 15:12.
[8]Heb. 12:1.
[9]Luke 15:10.
[10]Rev. 7:9.
[11]Rev. 21:16.
[12]Phil. 1:23; Heb. 10:34.
[13]I Cor. 2:9, 10.

My Response

1. My suggestions for next revision edition:

2. Send me information on:

    a. Melodyland Ministries   ( )

    b. Wills and trust confidential
       information   ( )

3. I have made a personal commitment to Jesus and believe the Lord died in my place. Because Jesus arose from the dead, I live. From here on, I confess Him as my Saviour.

Name_____
                    (please print)

Address_____

City_____State_____Zip_____

(Mail this and you will receive free literature on now to walk with God.)

Notes

Notes

Notes

Notes

Notes

Notes

Notes

Notes

Notes

Notes

Notes